CHEMISTRY PART 1

COMPLETE QUESTION BANK BOOK FOR CLASS 11

PRIYAMVADA UPADHYAY

ALL EXAM CRACKERS

Copyright © 2023 ALL EXAM CRACKERS

All rights reserved

The characters and events portrayed in this book are fictitious. Any similarity to real persons, living or dead, is coincidental and not intended by the author.

No part of this book may be reproduced, or stored in a retrieval system, or transmitted in any form or by any means, electronic, mechanical, photocopying, recording, or otherwise, without express written permission of the publisher.

ISBN : 9798850285395

Cover design by: Art Painter
Library of Congress Control Number: 2018675309
Printed in the United States of America

CONTENTS

Title Page	
Copyright	
SOME BASIC CONCEPTS OF CHEMISTRY	1
STRUCTURE OF AN ATOM	47
PERIODIC CLASSIFICATION OF ELEMENTS	87
Chemical Bonding and Molecular Structure	136
STATES OF MATTER	194
THERMODYNAMICS	212
EQUILIBRIUM	239
REDOX REACTION	273
HYDROGEN	299
THE S- BLOCK ELEMENTS	325
P - BLOCK ELEMENTS	356
Organic Chemistry – Some Basic Principles and Techniques	376
HYDROCARBONS	403
ENVIRONMENTAL CHEMISTRY	428
Books In This Series	457

SOME BASIC CONCEPTS OF CHEMISTRY

Question 1. Two students performed the same experiment separately, and each one of them recorded two separate readings of mass which are given below. The right reading of the mass is 3.0 g. Based on the provided data below, mark the correct option for the following statements.

Student Readings

(i)   (ii)
A   3.01   2.99
B   3.05   2.95

(i) Results of both the students are found neither accurate nor precise.

(ii) Results of student A are found both precise and accurate.

(iii) Results of student B are found neither precise nor accurate.

(iv) Student B's results are both precise and accurate.

Answer 1. Option (ii) is the correct answer.

Question 2. The measured temperature on the Fahrenheit scale is 200 °F. What would the reading

be on the Celsius scale?

(i) 40 °C

(ii) 94 °C

(iii) 93.3 °C

(iv) 30 °C

Answer 2. Option (iii) is the correct answer.

Question 3. What will be the molarity of the solution that contains 5.85 g of NaCl(s) per 500 mL?

(i) 4 mol L-1

(ii) 20 molL-1

(iii) 0.2 molL-1

(iv) 2molL-1

Answer 3. Option (iii) is the correct answer.

Question 4. When 500 mL of a given 5M solution is diluted to 1500 mL, what would be the molarity of the new solution?

(i) 1.5 M

(ii) 1.66 M

(iii) 0.017 M

(iv) 1.59 M

Answer 4. Option (ii) is the correct answer.

Question 5. The number of atoms in an element's mole equals the Avogadro number. Which among the following element contains the greatest number of atoms?

(i) 4g He

(ii) 0.40g Ca

(iii) 46g Na

(iv) 12g He

Answer 5. Option (iv) is the correct answer.

Question 6. When the glucose concentration ($C_6H_{12}O_6$) in the blood is 0.9 g L-1, what would be the molarity of glucose found in the blood?

(i) 0.5 M

(ii) 50 M

(iii) 0.005 M

(iv) 5 M

Answer 6. Option (iii) is the correct answer.

Question 7. What would be the molality for the solution obtained containing 18.25 g of HCl gas in 500 g of water?

(i) 0.5 m

(ii) 1 M

(iii) 0.1 m

(iv) 1 m

Answer 7. Option (iv) is the correct answer.

Question 8. One mole of any given substance contains 6.022 × 10^23 atoms/molecules. A number of molecules of H_2SO_4 that are present in 100 mL of 0.02M H_2SO_4 solution are _____.

(i) 12.044×10^{20} molecules

(ii) 12.044×10^{23} molecules

(iii) 1×10^{23} molecules

(iv) 6.022×10^{23} molecules

Answer 8. Option (i) is the correct answer.

Question 9. What is the mass percent of the carbon found in carbon dioxide?

(i) 0.034%

(ii) 27.27%

(iii) 3.4%

(iv) 28.7%

Answer 9. Option (ii) is the correct answer.

Question 10. The compound's empirical formula and molecular mass found are CH_2O and 180 g, respectively. What would be the molecular formula of the given compound?

(i) $C_9H_{18}O_9$

(ii) CH_2O

(iii) $C_6H_{12}O_6$

(iv) $C_2H_4O_2$

Answer 10. Option (iii) is the correct answer.

Question 11. When the density of the solution is 3.12 g mL-1, the mass of 1.5 mL solution in the significant figures is _____.

(i) 4.7g

(ii) 4680 × 10-3 g

(iii) 4.680g

(iv) 46.80g

Answer 11. Option (i) is the correct answer

Question 12. Which of the following statements given about the compound is incorrect?

(i) A molecule of the given compound has atoms of the different elements.

(ii) A compound can't be separated into its constituent elements through physical separation methods.

(iii) A compound retains the required physical properties of its constituent elements.

(iv) The ratio of atoms for the different elements in the compound is fixed.

Answer 12. Option (iii) is the correct answer.

Question 13. Which among the following statements is right about the reaction given below:

$4Fe(s) + 3O_2(g) \rightarrow 2Fe_2O_3(g)$

(i) The total mass of given iron and oxygen in reactants = total mass of iron and oxygen in the product; thus, it follows the law of conservation of mass.

(ii) The total mass of the given reactants = the total mass of product; thus, the law of multiple proportions followed.

(iii) Amount of Fe_2O_3 could increase by taking any of the reactants given (iron or oxygen) in excess.

(iv) Amount of Fe_2O_3 produced will decrease when the amount of any of the given reactants (iron as well as oxygen) taken is in excess.

Answer 13. Option (i) is the correct answer.

Question 14. Which among the following reactions is incorrect as per the law of conservation of the mass.

(i) $2Mg(s) + O_2(g) \rightarrow 2MgO(s)$

(ii) $C_3H_8(g) + O_2(g) \rightarrow CO_2(g) + H_2O(g)$

(iii) P4(s) + 5O2(g) → P4O10(s)

(iv) CH4(g) + 2O2(g) → CO2(g) + 2H2O (g)

Answer 14. Option (ii) is the correct answer.

Question 15. The following statements indicate the law of the multiple proportions.

(i) Sample of given carbon dioxide taken from any source will always have carbon and oxygen in the ratio of 1:2.

(ii) Carbon forms the two oxides, namely CO2 and CO. Here, masses of oxygen that combine with the fixed mass of carbon are in the simple ratio of 2:1.

(iii) When magnesium burns in the given oxygen, the amount of magnesium that took for the reaction is the same as the amount of magnesium found in the magnesium oxide formed.

(iv) At the given constant temperature and pressure, 200 mL of hydrogen would combine with 100 mL of oxygen to produce 200 mL of water vapour.

Answer 15. Option (ii) is the correct answer.

Question 16. One mole of oxygen gas formed at STP is equal to _____.

(i) 6.022 × 1023 molecules of the oxygen

(ii) 6.022 × 1023 atoms of the oxygen

(iii) 16 g of the oxygen

(iv) 32 g of the oxygen

Answer 16. Option (i) and (iv) are the correct answers.

Question 17. Sulphuric acid reacts with the given sodium hydroxide as follows as given below:

$H_2SO_4 + 2NaOH \rightarrow Na_2SO_4 + 2H_2O$

if 1L of 0.1M sulphuric acid solution is allowed to react with the given 1L of the 0.1M sodium hydroxide solution, the amount of the sodium sulphate formed will be, and its molarity in the solution got

(i) 0.1 mol L-1

(ii) 7.10 g

(iii) 0.025 mol L-1

(iv) 3.55 g

Answer 17. Option (ii) and (iii) are the right answers.

Question 18. Which among the following pairs has the same number of atoms?

(i) 16 g of O_2(g) as well as 4 g of H_2(g)

(ii) 16 g of O_2 as well as 44 g of CO_2

(iii) 28 g of N_2 as well as 32 g of O_2

(iv) 12 g of C(s) as well as 23 g of Na(s)

Answer 18. Option (iii) and (iv) are the right answers.

Question 19. Which among the following solutions has same concentration?

(i) 20 g of the NaOH in 200 mL of the solution

(ii) 0.5 mol of KCl in 200 mL of the solution

(iii) 40 g of NaOH in 100 mL of the solution

(iv) 20 g of KOH in 200 mL of the solution

Answer 19. Option (i) and (ii) are the right answers.

Question 20. 16 g of given oxygen has the same number of

molecules as in

(i) 16 g of CO

(ii) 28 g of N2

(iii) 14 g of N2

(iv) 1.0 g of H2

Answer 20. Option (iii) and (iv) are the right answers.

Question 21. Which among the following terms is unitless?

(i) Molality

(ii) Molarity

(iii) Mole fraction

(iv) Mass percent

Answer 21. Option (iii) and (iv) are the right answers.

Question 22. One of the following statements of Dalton's atomic theory given below:

"Compounds are formed for the following when atoms of different elements combine in a fixed ratio."

Which among the following laws is not related to this statement?

(i) Law of conservation of mass

(ii) Law of definite proportions

(iii) Law of multiple proportions

Answer 22. Option (i) and (iv) are the correct answers.

Question 23. What would be the mass of one atom of given C-12 in grams?

Answer 23. 1 mole of given carbon atom = 12g= 6.022×10^{23}

atoms.

Question 24. How many significant figures should be there to answer the following given calculations?

2.5 1.25 3.5/2.01

Answer 24. Two significant figures should be there in this.

As the least number of significant figures from the given figure is 2 (in 2.5 and 3.5).

Question 25. What is the symbol for the SI unit of the given mole? How can the mole be defined?

Answer 25. The symbol for the SI unit for the given mole is mol. The mole is defined as the amount of substance that contains as many entities as there are atoms in the given 12g carbon.

Question 26. What is the difference between the term molality and molarity?

Answer 26. Molarity is the provided number of moles of solute dissolved in 1 litre of the solution. Molality is the provided number of moles of solute present in 1kg of the solvent.

Question 27. Calculate the mass percent of the calcium, phosphorus and oxygen in the given calcium phosphate $Ca_3(PO_4)$

Answer 27. Molecular mass of the given $Ca_3(PO_4)$ = 3*40+2*31+8*16 =310

Mass percent of the given Ca = 3*40/310*100 = 38.71%

Mass percent of the given P = 2*31/310*100 = 20%

Mass percent of the given O = 8*16/310 = 41.29%

Question 28. 45.4 L of dinitrogen, when reacted with 22.7 L of the given dioxygen and 45.4 L of the nitrous oxide, was formed. The reaction is given below as:

$2N_2(g) + O_2(g) \rightarrow 2N_2O(g)$

Which law obey for the following experiment? Write the statement of the given law.

Answer 28. The above experiment proves the Gay-Lussac's law which states that the given gases combine or are produced in the chemical reaction in a simple whole-number ratio by volume, provided that all the given gases are at the same temperature and pressure.

Question 29. If two given elements can combine to form more than one compound, the masses of one element, which combines with the fixed mass of the other element, are in the whole-number ratio.

(a) Is this statement true?

(b) If yes, according to which given law?

(c) Give one example related to this law.

Answer 29. (a) Yes, the statement is true.

(b) As per the law of multiple proportions

(c), hydrogen and oxygen react to form given water and hydrogen peroxide

$H_2 + 1/2 O_2 \rightarrow H_2O$

$H_2 + O_2 \rightarrow H_2O_2$

Masses of oxygen which combine with the fixed mass of hydrogen are in the ratio of 16:32

Question 30. Calculate the average atomic mass of the given hydrogen using the following data :

Isotope	% Natural abundance	Molar mass
^1H	99.985	1

2H   0.015   2

Answer 30.

Average atomic mass is equal to

= 99.985*1+0.015*2/100

= 099.985*1+0.015*2/100

= 1.00015u

Question 31. Hydrogen gas is prepared in the laboratory by reacting the given diluted HCl with granulated zinc. The following reaction takes place.

$Zn + 2HCl \rightarrow ZnCl_2 + H_2$

Calculate the volume of the given hydrogen gas liberated at STP if 32.65 g of zinc reacts with HCl. 1 mol of the given gas occupies 22.7 L volume at STP; atomic mass of the Zn = 65.3 u.

Answer 31. 1 mol of the given gas occupies = 22.7L Volume at STP atomic mass of Zn = 65.3u

From the above-given equation,

65.3g of Zn if reacts with HCl produces = 22.7L H_2 at STP

Thus, 32.65g of Zn, when reacted with the given HCl, will produce

= 22.7 * 32.65/65.3

= 11.35L of H_2 at STP

Question 32. The density of 3 molal solutions of given NaOH is 1.110 g mL–1. Calculate the required molarity of the solution.

Answer 32. 3 molal solution of given NaOH = 3 moles of given NaOH dissolved in 1000g water

3 mole of given NaOH = 3*40g = 120g

Density of the solution = 1.110gmL-1

thus, Volume = mass/density = 1120g/1.110gmL-1 = 1.009L

Molarity of the given solution = 3/1.009 = 2.97M

Question 33. The volume of the solution changes with a change in the temperature. Then, would the molality of the solution be affected by the temperature? Give the reason for your answer.

Answer 33. Mass does not change because the temperature changes. Thus, the molality of the solution does not change.

Molality = moles of the solute/ weight of the solvent (in g) *1000

Question 34. If 4 g of the NaOH dissolves in 36 g of H_2O, calculate the mole fraction of each

component in the solution. Here, determine the molarity of the solution (specific

gravity of solution is 1g mL−1).

Answer 34. Mole fraction of the H_2O = number of moles of the H_2O/ Total no: of moles (H_2O+NaOH)

No: of moles of the H_2O = 36/18=2 moles

No: of moles of the NaOH = 4/40=0.1mol

Total number of the moles = 2+0.1= 2.1

Mole fraction of the H_2O = 2/2.1 = 0.952

Mole fraction of the NaOH = 0.1/2.1 = 0.048

Mass of the solution = Mass of H_2O + Mass of NaOH = 36+4=40G

Volume of the solution = 40/1 = 40mL

Molarity = 0.1/0.04 = 2.5M

Question 35. The reactant completely consumed in the reaction is called the limiting reagent.

In the reaction $2A + 4B \rightarrow 3C + 4D$, if 5 moles of A react with 6 moles of B,

then

(i) which is the limiting reagent?

(ii) calculate the amount of C formed?

Answer 35. (i) B would be the limiting reagent as it gives the lesser amount of product.

(ii) Let B completely consumed

4 mol of B gives 3 mol of C

6 mol of B will give 3/4 *6 mol C = 4.5 mol C

Question 36. Match The Following Type

(i) 88 g of CO_2
(ii) 6.022×10^{23} molecules of the H_2O

(iii) 5.6 litres of O_2 at STP

(iv) 96 g of O_2

(v) 1 mol for any gas

(a) 0.25 mol
(b) 2 mol

(c) 1 mol

(d) 6.022×10^{23} molecules

(e) 3 mol

Answer 36. A → b

B → c

C → a

D → e

E → d

Question 37. Match the following

Physical quantity   Unit
(i) Molarity
(ii) Mole fraction

(iii) Mole

(iv) Molality

(v) Pressure

(vi) Luminous intensity

(vii) Density

(viii) Mass

(a) g mL–1
(b) mol

(c) Pascal

(d) Unitless

(e) mol L–1

(f) Candela

(g) mol kg–1

(h) Nm–1

(i) kg

Answer 37. (i → e)

(ii → d)

(iii → b)

(iv → g)

(v → c)

(vi → f)

(vii → a)

(viii → i)

Question 38. Assertion (A): Ethene's empirical mass is half its molecular mass.

Reason (R): The empirical formula presents the simplest whole-number

the ratio of various atoms present in the compound.

(i) Both A as well as R are true, and R is the correct explanation of A.

(ii) A is true, as well as R is false.

(iii) A is false, as well, as R is true.

(iv) Both A as well as R are false.

Answer 38. Option (i) is correct.

Question 39. Assertion (A): One atomic mass unit has defined as one-twelfth of the mass of

one carbon-12 atom.

Reason (R): Carbon-12 isotope is the most abundant isotope of carbon

and has been chosen as the standard.

(i) Both A and R are true, and R is the correct explanation of A.

(ii) Both A and R are true, but R is not the correct explanation of A.

(iii) A is true, but R is false.

(iv) Both A and R are false.

Answer 39. Option (ii) is correct. Carbon-12 is considered a standard for defining atomic and molecular mass.

Question 40. Assertion (A): Significant figures for the 0.200 is 3 here; for 200 it is 1.

Reason (R): Zero at the end or the right of the number is significant, provided

they are not on the right side for the decimal point.

(i) Both A and R are true, as well as R is the correct explanation of A.

(ii) Both A and R are true, as well as R is not a correct explanation of A.

(iii) A is true, as well as R is false.

(iv) Both A as well as R are false.

Answer 40. Option (iii) is correct. Significant figures for the 0.200 = 3 and for 200 = 1

Zero at the end of the number without a decimal point may or may not be significant depending upon the accuracy of the measurement.

Question 41. Assertion (A): Combusting 16 g of methane gives 18 g of water.

Reason (R): In the combustion of methane, water is one of the products.

(i) Both A and R are true, but R is not the correct explanation of A.

(ii) A is true, but R is false.

(iii) A is false, but R is true.

(iv) Both A and R are false.

Answer 41. Option (iii) is correct.

16g of CH4 on complete combustion will give 36g of water.

Question 42. A vessel contains 1.6 g of dioxygen at STP (273.15K, 1 atm pressure). The

gas transfers to another vessel at a constant temperature, where

the pressure becomes half of the original pressure. Calculate

(i) the volume of the new vessel.

(ii) a number of molecules of dioxygen.

Answer 42. (i) Moles of the oxygen = 1.6/32 = 0.05mol

At STP, 1 mol of the O2 = 22.4L

now, volume of O2 = 22.4 × 0.05 = 1.12L

V1 = 1.12L

V2 =?

P1 = 1atm

P2 = ½ = 0.5atm

As per, Boyle's law, P1V1 = P2V2

thus, Substituting the value

V2 = 1 × 1.12/0.5 = 2.24L

(ii) Number of molecules in the 1.6g or 0.005mol = 6.022 × 10^{23} × 0.05 = 3.011 × 10^{22}

Question 43. Calcium carbonate reacts with the aqueous HCl to give $CaCl_2$ and CO_2 according

to the reaction given below:

$CaCO_3$ (s) + 2HCl (aq) → $CaCl_2$(aq) + CO_2(g) + H_2O(l)

What would be form masses of $CaCl_2$ if 250 mL of 0.76 M HCl reacts with

1000 g of the $CaCO_3$? Name the limiting reagent. Calculate the number of the moles

of $CaCl_2$ formed in the reaction.

Answer 43. No: of moles of the HCl taken = MV/1000 = 0.76*250/1000 = 0.19

No: of moles of the $CaCO_3$ = Mass/Molar mass = 1000/100 = 10

if $CaCO_3$ is completely consumed
1 mol of $CaCO_3$ = 1 mol $CaCl_2$

10 mol $CaCO_3$ = 10mol $CaCl_2$

if HCl will consume completely.
2 mol HCl = 1 mol $CaCl_2$

0.19mol of HCl = ½ × 0.19mol $CaCl_2$ = 0.095 mol $CaCl_2$

HCl would be the limiting reagent, and the number of moles of $CaCl_2$ formed will be 0.095mol

Question 44. A box contains a few identical red coloured balls that have to label as A, each weighing

2 grams. Another box contains an identical blue coloured ball and labelled as B,

each weighing 5 grams. Consider the combinations of AB, AB2, A2B and A2B3 and show that the law of the multiple proportions is applicable.

Answer 44.

	AB	AB2	A2B	A2B3
Mass of the A (in g)	2	2	4	4
Mass of the B (in g)	5	10	5	15

Masses of B, which combine with a fixed mass of A, are

10g, 20g, 5g, 15g

2: 4: 1 : 3

This is the simple whole-number ratio.

Question 45. In a reaction

A + B2 → AB2

Identify the limiting reagent, if any, in the following reaction mixtures.

(i) 300 atoms of A + 200 molecules of B

(ii) 2 mol of A + 3 mol of B

(iii) 100 atoms of A + 100 molecules of B

(iv) 5 mol of A + 2.5 mol of B

(v) 2.5 mol of A + 5 mol of B

Answer 45. Limiting reagent:

It determines the extent of a reaction. It is the first to get consumed during a reaction, thus causing the reaction to stop and limiting the amount of product formed.

(i) 300 atoms of A + 200 molecules of B

1 atom of A reacts with 1 molecule of B. Similarly, 200 atoms of A react with 200 molecules of B, so 100 atoms of A are unused. Hence, B is the limiting reagent.

(ii) 2 mol A + 3 mol B

1 mole of A reacts with 1 mole of B. Similarly, 2 moles of A react with 2 moles of B, so 1 mole of B is unused. Hence, A is the limiting reagent.

(iii) 100 atoms of A + 100 molecules of Y

1 atom of A reacts with 1 molecule of Y. Similarly, 100 atoms of A react with 100 molecules of Y. Hence, it is a stoichiometric mixture where there is no limiting reagent.

(iv) 5 mol of A + 2.5 mol of B

1 mole of A reacts with 1 mole of B. Similarly, 2.5 moles of A react with 2.5 moles of B, so 2.5 moles of A are unused. Hence, B is the limiting reagent.

(v) 2.5 mol of A + 5 mol of B

1 mole of A reacts with 1 mole of B. Similarly, 2.5 moles of A react with 2.5 moles of B, so 2.5 moles of B are unused. Hence, A is the limiting reagent.

Question 46. Define the law of multiple proportions. Explain it with the two examples. How does this law point to the existence of atoms?

Answer 46. Dalton first studied the law of multiple proportions in 1803, and it can be found and follows.

If two elements combine to form two or more chemical compounds, the masses of one of the elements combine with the fixed mass of the other in a simple ratio.

For example, hydrogen reacts with oxygen to form two compounds: water and hydrogen peroxide.

Hydrogen(2g) + Oxygen(16g) → H2O(18g)

Hydrogen(2g) + Oxygen(32g) → H2O2(34g)

In this case, the masses of the oxygen (i.e. 16g and 32g) that combine with the fixed mass of hydrogen (2g) have a simple ratio, i.e. 16:32 or 1:2.

As we all know, if compounds mix in different proportions, they form different compounds. For example, if hydrogen mixes with a different proportion of oxygen, it forms water or hydrogen peroxide.

It demonstrates that there are constituents that combine in a specific manner. This constituent could be atoms. As a result, the law of multiple proportions demonstrates the existence of atoms that combine to form the molecules.

Question 47. A box contains a few identical red coloured balls, labelled as A, each weighing equal to 2 grams. The other box contains identical blue coloured balls, labelled for the following as B, each weighing 5 grams. Consider the following combinations AB, AB2, A2B and A2B3, and show that the multiple proportions law applies.

Answer 47.

Combination	Mass of A (g)	Mass of B (g)
AB	2	5
AB2	2	10
A2B	4	5
A2B3	4	15

When the two elements combine to form the given two or more compounds, the different masses of one element, which combine with a fixed mass of the other, bear a simple ratio to one another, as per the law of multiple proportions.

The mass of B, if combined with the fixed mass of A (say 1g), is 2.5g, 5g, 1.25g, and 3.75g. They have a 2:4:1 ratio, which is a simple whole-number ratio. Hence. The multiple proportions law is applicable.

Question 48. Assertion (A): an atomic mass unit defined as one-twelfth of the mass of a given one carbon-12 atom.

Reason (R): Carbon-12 isotope has the most abundant isotope of the given carbon and choice as standard.

(i) Both A as well as R are true, and R is the correct explanation of A.

(ii) Both A as well as R are true, but R is not the correct explanation for A.

(iii) A is true, as well as R is false.

(iv) Both A and R are false.

Answer 48. The correct option is (i) Both A and R are true, and R is the correct explanation of A.

As C-12 is used as the standard atom, one atomic mass unit is defined as one-twelfth of the mass of the given one carbon – 12 atom. That is because it has an equal number of protons and neutrons (6) and forms the majority of matter.

Carbon-12 is the most abundant isotope of all carbon's isotopes.

Question 49. Assertion (A): Significant figures for the number 0.200 are 3, whereas, for the number 200, it is 1.

Reason (R): Zero at the end or right of the number is significant, provided they don't give on the correct side of the decimal point.

(1) Both A as well as R are true, and R is the correct explanation for A.

(ii) Both A as well as R are true, but R is not the correct explanation for A.

(iii) A is true, as well as R is false.

(iv) Both A as well as R are false.

Answer 49. The correct option is (iii) A is true, as well as R is false.

Zero at the end, as well as to the right of the number, is significant when this is on the correct side of the decimal point. For example, 0.200 has 3 significant figures.

Question 50. Assertion (A): Combustion of 16 g for methane gives 18 g of water.

Reason (R): Water is one of the products in the combustion of methane.

(i) Both A as well as R are true, but R is not the correct explanation for A.

(ii) A is true, as well as R is false.

(iii) A is false, as well, as R is true.

(iv) Both A as well as R are false.

Answer 50. The correct option is (iii) A is false, but R is true.

$CH_4 + O_2 \rightarrow CO_2 + 2H_2O$

Water produces during the combustion for methane, but 16 g of methane on complete combustion gives 36 g of water.

Question 51. Calculate the molecular mass of the following:

(i) H_2O

(ii) CO_2

(iii) CH_4

Answer 51. (i) CH_4 :

The molecular weight of methane, CH_4

= (1 x Atomic weight of carbon) + (4 x Atomic weight of hydrogen)

= [1(12.011 u) +4 (1.008u)]

= 12.011u + 4.032 u

= 16.043 u

(ii) H_2O :

The molecular weight of water, H_2O

= (2 x Atomic weight of hydrogen) + (1 x Atomic weight of oxygen)

= [2(1.0084) + 1(16.00 u)]

= 2.016 u +16.00 u

= 18.016u

So approximately

= 18.02 u

(iii) CO_2 :

= Molecular weight of carbon dioxide, CO_2

= (1 x Atomic weight of carbon) + (2 x Atomic weight of oxygen)

= [1(12.011 u) + 2(16.00 u)]

= 12.011 u +32.00 u

= 44.011 u

So approximately

= 44.01u

Question 52. The following data obtained if dinitrogen and

dioxygen react together to form different compounds:

Mass of dinitrogen	Mass of dioxygen
(i) 14 g	16 g
(ii) 14 g	32 g
(iii) 28 g	32 g
(iv) 28 g	80 g

(a) Which chemical combination law obeys the above experimental data? Given its statement.

(b) Fill in the blanks in the following conversions given below:

(i) 1 km = ... mm = ... pm

(ii) 1 mg = ... kg = ... ng

(iii) 1 mL = ... L = ... dm^3

Answer 52. (a) Fixing the mass of the dinitrogen as 28 g, masses of the dioxygen combined would be 32, 64, 32 and 80 g in given four oxides. These masses of the dioxygen bear a simple whole number ratio as 2:4:2:5. Therefore, the data given will obey the law of the multiple proportions.

The statement is as follows, two elements always combine in the fixed mass of another, bearing a simple ratio to another to form the two or more chemical compounds.

(b) (i) 1 km = 1km × 1000m/ 1km × 100cm /1m/ 10mm /1cm = 10^6 mm

1 km = 1km × 1000m / 1km × 1pm/ 10^{-12} m = 10^{15} pm

(ii) 1 mg = 1mg × 1g/ 1000mg × 1kg / 1000g = 10^{-6} kg

1 mg = 1mg × 1g/ 1000mg × 1ng/ 10^{-9}g = 10^{-6} ng

(iii) 1 mL = 1mL × 1L/ 1000mL = 10^{-3} L

$1 \text{ mL} = 1\text{cm}^3 = 1\text{cm}^3 \times (1\text{dm} \times 1\text{dm} \times 1\text{dm}/ 10\text{cm} \times 10\text{cm} \times 10\text{cm}) = 10^3 \text{dm}^3$.

Question 53. When the speed of light is $3.0 \times 10^8 \text{ ms}^{-1}$, calculate the distance for the following covered by the light in 2.00 ns.

Answer 53. Distance covered will be

= Speed × Time = $3.0 \times 10^8 \text{ ms}^{-1} \times 2.00 \text{ ns}$

= $3.0 \times 10^8 \text{ ms}^{-1} \times 2.00 \text{ ns} \times 10^{-9} \text{ s} /1\text{ns}$ = 6.00×10^{-1} m = 0.600 m

Question 54. When ten volumes of dihydrogen gas react with five volumes of dioxygen gas, how many volumes of water vapour will produce?

Answer 54. Dihydrogen gas reacts with the given dioxygen gas as,

$2H_2(g) + O_2(g) \rightarrow 2H_2O(g)$

Therefore, the two volumes of dihydrogen react with one volume of dihydrogen to produce two volumes of water vapour. So, the ten volumes of dihydrogen would react with the five volumes of dioxygen to produce the required ten volumes of water vapour.

SET 2

1. Elements X and Y combine to form two compounds XY and X_2Y. Find the atomic weight of X and Y, if the weight of 0.1 moles of XY is 10g and 0.05 moles of X_2Y is 9g

(a) 30, 20

(b) 80, 20

(c) 60, 40

(d) 20, 30

Answer: (b)

2. Which one will have maximum numbers of water molecules?

(a) 18 molecules of water

(b) 1.8 grams of water

(c) 18 grams of water

(d) 18 moles of water

Answer: (d)

3. The number of atoms present in 0.1 moles of a triatomic gas is

(a) 1.806×10^{23}

(b) 1.806×10^{22}

(c) 3.600×10^{23}

(d) 6.026×10^{22}

Answer: (a)

4. Find the volume of O_2 required to burn 1 L of propane completely, measured at 0°C temperature and 1 atm pressure

(a) 10 L

(b) 7 L

(c) 6 L

(d) 5 L

Answer: (d)

5. A gas X has C_p and C_v ratio as 1.4, at NTP 11.2 L of gas X will contain_____ number of atoms

(a) 1.2×10^{23}

(b) 3.01×10^{23}

(c) 2.01×10^{23}

(d) 6.02×10^{23}

Answer: (d)

6. Which of the species is not

paramagnetic?

(a) As^+

(b) Cl^-

(c) Ne^{2+}

(d) Be^+

Answer: (b)

7. Pressure has the same dimension as _____

(a) energy per unit volume

(b) energy

(c) force per unit volume

(d) force

Answer: (a)

8. A container has an equal mass of H_2, O_2 and CH_4 at 27°C, the ratio of their volume is

(a) 16:8:1

(b) 8:1:2

(c) 16:1:2

(d) 8:16:1

Answer: (c)

9. There are two chlorides of sulphur S_2Cl_2 and SCl_2. What is the equivalent mass of S in SCl_2

(a) 64.8 g/mole

(b) 32 g/mole

(c) 16 g/mole

(d) 8 g/mole

Answer: (c)

10. Boron exists as two stable isotopes; $^{10}B(19\%)$ and $^{11}B(81\%)$. Find out the average atomic weight of boron in the periodic table

(a) 10.0

(b) 11.2

(c) 10.2

(d) 10.8

Answer: (d)

11. The mass of carbon-12 atom considered in the definition of a mole is

(a) 0.12g

(b) 0.012 kg

(c) 120 mg

(d) None of these

Answer: (b)

12. Which of the following is a pair of physical and chemical property respectively of a substance

(a) density and acidity

(b) basicity and colour

(c) colour and density

(d) acidity and combustibility

Answer: (a)

13. The normality of 0.3M phosphorous acid is

(a) 0.9

(b) 0.1

(c) 0.6

(d) 0.3

Answer: (c)

14. 3g of an oxide of a metal is converted into chloride and it yielded 5g of chloride. Find the equivalent weight of the metal

(a) 20

(b) 12

(c) 3.325

(d) 33.25

Answer: (d)

15. 0.32 gm of a metal on treatment with an acid gave 112mL of hydrogen at

STP. Calculate the equivalent weight of the metal.

(a) 24

(b) 11.2

(c) 32

(d) 58

Answer: (c)

16. Which of the following is not a homogenous mixture?

(a) Brass

(b) Air

(c) Smoke

(d) Aqueous solution of sugar

Answer: (c)

COMPETITIVE CORNER

NEET

Q: Given the number, 161 cm, 0.161 cm, 0.0161 cm. The number of significant figures for the three numbers is: [AIPMT 1998]

A) 3, 4 and 5, respectively

B) 3, 4 and 4, respectively

C) 3, 3 and 4, respectively

D) 3, 3 and 3, respectively

ANS: D

Q: Haemoglobin contains 0.33% of iron by

weight. The molecular weight of haemoglobin is approximately 67200. The number of iron atoms (at. wt. of Fe is 56) present in one molecule of haemoglobin are: [AIPMT 1998]

A) 1

B) 6

C) 4

D) 2

ANS:C

Q: An organic compound containing C, H and N gave the following results on analysis C = 40%, H = 13.33%, N = 46.67%. Its empirical formula would be: [AIPMT 1998]

A) **C2H7N2**

B) **CH5N**

C) **CH4N**

D) **C2H7N**

Correct Answer: C

Q *1.0 g of magnesium is burnt with 0.56 g O2 in a closed verssel. Which reactant is left in excess and how much?   [AIPMT 2014]

(At.wt.Mg=24;O=16)

A) Mg,0.16g

B) O2.0.16g

C)Mg,0.44

D) $O_2, 0.28g$

Correct Answer: A

Q: If Avogadro number N_A, is changed from $6.022 \times 10^{23} mol^{-1}$ to $6.022 \times 10^{20} mol^{-1}$ this would change [NEET 2015 (Re)]

A) the definition of mass in units of grams

B) the mass of one mole of carbon

C) the ratio of chemical species to each other in a balanced equation

D) the ratio of elements to each other in a compound

Correct Answer: A

Q: **Which of the following is dependent on temperature?**

[NEET-2017]

A) Weight percentage

B)   Molality

C) Molarity

D)   Mole fraction

Correct Answer: C

Q: In which case is number of molecules of water maximum? [NEET - 2018]

A) 0.00224 L of water vapours

at 1 atm and 273 K

B) 0.18 g of water

C) 18 mL of water

D) 10--3 mol

of water

Correct Answer: C

Q: **Which one of the followings has maximum number of atoms?** [NEET 2020]

A) 1 g of Mg(s) [Atomic mass of Mg = 24]

B) 1 g of O2(g) [Atomic mass of O = 16]

C) 1 g of Li(s) [Atomic mass of Li = 7]

D) 1 g of Ag(s) [Atomic mass of Ag = 108]

Correct Answer: C

1. Elements X and Y combine to form two compounds XY and X_2Y. Find the atomic weight of X and Y, if the weight of 0.1 moles of XY is 10g and 0.05 moles of X_2Y is 9g

(a) 30, 20

(b) 80, 20

(c) 60, 40

(d) 20, 30

Answer: (b)

2. Which one will have maximum numbers of water molecules?

(a) 18 molecules of water

(b) 1.8 grams of water

(c) 18 grams of water

(d) 18 moles of water

Answer: (d)

3. The number of atoms present in 0.1 moles of a triatomic gas is

(a) 1.806×10^{23}

(b) 1.806×10^{22}

(c) 3.600×10^{23}

(d) 6.026×10^{22}

Answer: (a)

4. Find the volume of O_2 required to burn 1 L of propane completely, measured at 0°C temperature and 1 atm pressure

(a) 10 L

(b) 7 L

(c) 6 L

(d) 5 L

Answer: (d)

5. A gas X has C_p and C_v ratio as 1.4, at NTP 11.2 L of gas X will contain_____ number of atoms

(a) 1.2×10^{23}

(b) 3.01×10^{23}

(c) 2.01×10^{23}

(d) 6.02×10^{23}

Answer: (d)

6. Which of the species is not paramagnetic?

(a) As^+

(b) Cl^-

(c) Ne^{2+}

(d) Be^+

Answer: (b)

7. Pressure has the same dimension as _____

(a) energy per unit volume

(b) energy

(c) force per unit volume

(d) force

Answer: (a)

8. A container has an equal mass of H_2, O_2 and CH_4 at 27°C, the ratio of their volume is

(a) 16:8:1

(b) 8:1:2

(c) 16:1:2

(d) 8:16:1

Answer: (c)

9. There are two chlorides of sulphur S_2Cl_2 and SCl_2. What is the equivalent mass of S in SCl_2

(a) 64.8 g/mole

(b) 32 g/mole

(c) 16 g/mole

(d) 8 g/mole

Answer: (c)

10. Boron exists as two stable isotopes; $^{10}B(19\%)$ and $^{11}B(81\%)$. Find out the average atomic weight of boron in the periodic table

(a) 10.0

(b) 11.2

(c) 10.2

(d) 10.8

Answer: (d)

11. The mass of carbon-12 atom considered in the definition of a mole is

(a) 0.12g

(b) 0.012 kg

(c) 120 mg

(d) None of these

Answer: (b)

12. Which of the following is a pair of physical and chemical property respectively of a substance

(a) density and acidity

(b) basicity and colour

(c) colour and density

(d) acidity and combustibility

Answer: (a)

13. The normality of 0.3M phosphorous acid is

(a) 0.9

(b) 0.1

(c) 0.6

(d) 0.3

Answer: (c)

14. 3g of an oxide of a metal is converted into chloride and it yielded 5g of chloride. Find the equivalent weight of the metal

(a) 20

(b) 12

(c) 3.325

(d) 33.25

Answer: (d)

15. 0.32 gm of a metal on treatment with an acid gave 112mL of hydrogen at STP. Calculate the equivalent weight of the metal.

(a) 24

(b) 11.2

(c) 32

(d) 58

Answer: (c)

16. Which of the following is not a homogenous mixture?

(a) Brass

(b) Air

(c) Smoke

(d) Aqueous solution of sugar

Answer: (c)

STRUCTURE OF AN ATOM

Question 1. Which among the following conclusions could not be derived from Rutherford's α -particle scattering experiment?

(i) Most of the given space in the atom is empty.

(ii) The atom's radius is about 10–10 m, while that of the nucleus is 10–15 m.

(iii) Electrons move in the circular path of the fixed energy called orbits.

(iv) Electrons and the nucleus are held together by electrostatic forces of attraction.

Answer 1. Option (iii) is the correct answer.

Question 2. Which options don't represent an atom's ground state electronic configuration?

(i) 1s2 2s2 2p6 3s2 3p6 3d8 4s2

(ii) 1s2 2s2 2p6 3s2 3p6 3d9 4s2

(iii) 1s2 2s2 2p6 3s2 3p6 3d10 4s1

(iv) 1s2 2s2 2p6 3s2 3p6 3d5 4s1

Answer 2. Option (ii) is the correct answer.

Question 3. The probability density plots of the 1s and 2s

orbitals have given in Fig.

The density of dots in the region represents the probability density of finding an electron in the region.

Based on the above diagram, which among the following statements is incorrect?

(i) 1s and 2s orbitals are given as spherical.

(ii) The probability of the finding the electron is close to the nucleus.

(iii) probability of finding the electron at the given distance is equal in all directions.

(iv) probability density of electrons for 2s orbital decreases uniformly and gradually as the distance from the nucleus increases.

Answer 3. Option (iv) is the correct answer.

Question 4. Which of the following statements will not correct the characteristics of cathode rays?

(i) They start from the cathode and then move towards the anode.

(ii) They travel in a straight line without an external electrical or magnetic field.

(iii) Characteristics of cathode rays don't depend upon the material of electrodes in the cathode ray tube.

(iv) Characteristics of cathode rays depend entirely upon the nature of gas present in the cathode ray tube.

Answer 4. Option (iv) is the correct answer.

Question 5. Which among the following statements about the electron is incorrect?

(i) It is the negatively charged particle.

(ii) The mass of the electron is equal to the mass of the neutron.

(iii) It is the primary constituent of all atoms.

(iv) It is the constituent of cathode rays.

Answer 5. Option (ii) is the correct answer.

Question 6. Which of the following atom properties could be explained correctly by the Thomson Model of an atom?

(i) Overall neutrality of the atoms.

(ii) Spectra of the hydrogen atom.

(iii) Position of the atom's electrons, protons and neutrons.

(iv) Stability of the atoms.

Answer 6. Option (i) is the correct answer.

Question 7. Two atoms are said to be isobars only if.

(i) they have the same atomic number but different mass numbers.

(ii) they have the equal number of electrons but different numbers of neutrons.

(iii) they have the equal number of neutrons but different numbers of electrons.

(iv) the addition of the number of protons and neutrons is equal, but the number of Protons is different.

Answer 7. Option (iv) is the correct answer

Question 8. The number of the required radial nodes for 3p orbital is _____.

(i) 3

(ii) 4

(iii) 2

(iv) 1

Answer 8. Option (iv) is the correct answer.

Question 9. Number of the required angular nodes for 4d orbital is _____.

(i) 4

(ii) 3

(iii) 2

(iv) 1

Answer 9. Option (iii) is the correct answer.

Question 10. Which among the following is responsible for ruling out the existence of definite paths or trajectories of the electrons?

(i) Pauli's exclusion principle.

(ii) Heisenberg's uncertainty principle.

(iii) Hund's rule of maximum multiplicity.

(iv) Aufbau principle.

Answer 10. Option (ii) is the correct answer.

Question 11. Arrange s, p and d sub-shells of the shell in the increasing order of their effective nuclear charge (Zeff) experienced by the electron present in them.

Answer 11. d<p<s

S orbital shields the electrons from the nucleus more than the p-orbitals shield more in d.

Question 12. Using the orbital diagram, show the distribution of the electrons in the oxygen atom (atomic number 8).

Answer 12. $_8O$ = 1s2 2s2 2p4

Question 13. The electronic configuration of the valence shell for Cu is 3d10 4s1 and not 3d94s2. How is this configuration explained?

Answer 13. Configuration with a filled and half-filled orbital has extra stability. In 3d104s1, d orbitals are completely filled and s orbitals are half-filled.

Question 14. The hydrogen atom possesses only one electron. Hence the mutual repulsion between the electrons is absent. But, in multielectron atoms, the mutual repulsion between the electrons is significant. How does this affect the energy of the given electron in the orbitals of the same principal quantum number in multielectron atoms?

Answer 14. The hydrogen atom possesses only one electron. Thus, the mutual repulsion between electrons is absent. But in the multielectron atoms, the mutual repulsion between the electrons is significant. Incomplete answer

Question 15. Out of electrons and protons, which one will have a higher velocity to produce matter waves of the same wavelength? Explain it.

Answer 15. Out of the electron and proton, being the lighter particle, the electron will have the higher velocity and produce matter waves of the same wavelength.

Question 16. Calculate the total number of required angular nodes and radial nodes present in the 3p orbital.

Answer 16. Nodes are the region present among the particular orbitals where one can find the probability density of finding electrons would be zero.

In case of np orbitals, the radial nodes,

= n − l − 1

= 3 − 1 − 1 = 1

Angular nodes = l = 1.

Question 17. A nickel atom can lose two electrons to form a Ni2+ ion. The atomic number of element nickel is 28. From which orbital would the nickel lose two electrons.

Answer 17. One Ni atom contains 28 electrons and the electronic configuration is : [Ar] 4s2 3d8

It becomes Ni2+ by losing two given electrons, so the configuration of Ni2+ is : [Ar] 4s0 3d8

Hence, nickel loses two electrons from the 4s orbital, not the 3d orbital, according to the Aufbau principle.

Question 18. Which among the following orbitals are degenerate?

3dxy, 4dxy, 3dz2, 3dyx, 4dyx, 4dzz

Answer 18. The energy of the orbitals depend upon the principal quantum number or the main shell to a large extent. Thus, the orbitals with the equal value of n will have the same energy levels and will be known as the degenerate orbitals.

Degenerate orbitals are 3dxy, 3dz2, and 3dyx as they have the same main shell n = 3.

Also, 4dxy, 4dyx, and 4dzz as they have the same value of n=4.

Question 19. An atom that has atomic mass number 13 has seven neutrons. What is the atomic number of the given atom?

Answer 19. The mass number of the atom = number of protons + number of neutrons

Hence, the atomic number (number of protons) = mass number – no. Of neutrons.

The atomic number of the given atom:

= 13 – 7 = 6.

Question 20. According to de Broglie, the matter would exhibit dual behaviour, that is, both particle and wave-like properties. However, a cricket ball with a mass of 100 g doesn't move like a wave when a bowler throws it at the speed of 100 km/h. Calculate the ball's wavelength and explain why it does not show wave nature.

Answer 20. m= 100g or 0.1kg

v= 100km/h =100*1000/60*60 = 1000/36m/s

thus, λ =h/mv = 2.387*10-34 m

Question 21. What experimental evidence supports the idea that an atom's electronic energies are quantised?

Answer 21. The bright-line spectrum shows that the energy level in an atom is quantised. These lines obtained the result of electronic transition between the energy and if the electronic energy level were continuous and not quantised or discrete; the atomic spectra would have shown a constant absorption(from the lower to higher energy level transition) or emission (from higher to lower energy level transition.

Question 22. What is the difference between terms orbit and orbital?

Answer 22. The orbit is the definite circular path for the electrons to revolve around the nucleus. It represents the two-dimensional motion of electrons around the nucleus. The orbital is not a well-defined path because it's a region around the nucleus where the probability of finding an electron is maximum.

Question 23. A hypothetical electromagnetic wave is shown in Fig. Find out the wavelength of the radiation.

Answer 23. wavelength defined as the distance between two successive alike points in a wave (usually between two maxima s, i.e. peaks, as well as two minima s, i.e. troughs, as shown in Fig.)

thus, for the given hypothetical wave, wavelength,

$\lambda = 4 * 2.16$ pm

$= 8.64$ pm.

Question 24. Which among the following will not show deflection from the path on passing through the electric field?

Proton, cathode rays, electron, neutron

Answer 24. Neutron would not show deflection from the path on passing through the electric field.

That is because of the neutral nature of the neutron particles. Hence, it has no charge and is not affected by an electric field.

For the following three particles, proton (positive), electron (negative), and cathode rays (the beam of the electrons, negatively charged) all have charges in them; thus, they will get deflected easily by an electric field.

Question 25. Chlorophyll present in green leaves of the plants absorbs light at 4.620×10^{14} Hz. Calculate the wavelength of the radiation in nanometers. Which part of the electromagnetic spectrum does it belong to?

Answer 25. Relation between wavelength as well as frequency can be expressed as :

$\lambda = c/v$, here c shows the velocity of light as well, and v shows the frequency for the radiation.

For given problem $\lambda = 3 \times 10^8$ ms^{-1} / 4.620×10^{14} Hz

$= 0.6494$ times 10^{-6} m^{-1}

Question 26. The Balmer series in the hydrogen spectrum corresponds to the transition from $n_1 = 2$ to $n_2 = 3, 4, \ldots$ This series lies in the visible region. Calculate the wave number of lines associated by the transition in the Balmer series if the electron moves to $n = 4$ orbit. ($R_H = 109677$ cm^{-1})

Answer 26. According to the Bohr's model for the hydrogen atom

$v = R_H(1/n_1^2 - 1/n_2^2)$ cm^{-1}

where, $n_1 = 2$ and $n_2 = 4$ and H = Rydberg's constant = 109677

therefore, wave number $-v = 109677 (¼ - 1/16)$

$= 20564.44$ cm^{-1}

Question 27. The effect of the uncertainty principle is quite significant only for the motion of microscopic particles and is negligible for the given macroscopic particles. Explain the statement with the help of a suitable example.

Answer 27. The uncertainty principle is only significantly applicable for the microscopic particles, not macroscopic particles. That concluded from the measurement of uncertainty:

For example, if we take the particle or an object of mass 1 milligram that is 10^{-6} kg

We calculate the following,

$\Delta x \cdot \Delta v = 60626 \ast 10^{-34} / 4 \ast 3.14 \ast 10^{-6}$

$= 10^{-28}$ m^{-2} s^{-1}

The value we get in this case is negligible and very insignificant for the concept of the uncertainty principle to apply to the particles.

Question 28. Table-tennis ball has a mass of 10 g and a speed of 90 m/s. If speed could be measured with the accuracy of 4%, what will be the uncertainty in speed and position?

Answer 28. According to Heisenberg's uncertainty principle :

"It is fundamentally impossible to accurately determine the velocity and position of a particle at the same time.

$\Delta x \cdot \Delta p \geq h/4\pi$

From given problem,

mass of the ball = 4 g and speed is = 90 m /s

hence, the uncertainty of speed shows as $\Delta v = 4/100 \times 90 = 3.6$ m/s

Δx has given by $\Delta x = h/4\pi m \Delta v$

Hence, the uncertainty of position shows as $\Delta x = 6.26 \times 10^{-34} / 4 \times 3.14 \times 4 \times 3.6$

$= 1.46 \times 10^{-33}$ m

Question 29. Wavelengths of the following different radiations have given below :

$\lambda(A) = 300$ nm

$\lambda(B) = 300$ μm

$\lambda(c) = 3$ nm

$\lambda(D)\ 30\ A°$

Arrange the following radiations in the increasing order of their energies.

Answer 29. As per the Planck's quantum theory, energy is related to the frequency of the given radiation by :

E = h × Frequency

So, E is proportional to $1/\lambda$

So, the relation b/w energy and wavelength are inversely proportional. Thus, the lower the wavelength higher would be the energy of the radiation.

For the following given wavelength

$\lambda(A)$ = 300 nm = 300 x 10-9 m = 3 x 10 -7 m

$\lambda(B)$ = 300 μm = 3 00 x 10-6 = 3 x 10-4

$\lambda(c)$ = 3 nm = 3 x 10 – 9

$\lambda(D)$ 30 A° = 3 X 10 – 9

The increment order of the following given wavelengths: $\lambda(c)$ = $\lambda(D) < \lambda(A) < \lambda(B)$

So, the increasing order of the given energy would be the opposite: $\lambda(B) < \lambda(A: < \lambda(c) = \lambda(D)$

Question 30. The arrangement of the orbitals is based upon the energy based on the (n+l) value. The lower the value of the (n+l), the lower energy found. For orbitals with equal values of (n+l), the orbital with the lower value of n will have the lower energy.

Based upon the above-given information, arrange the following orbitals in the increasing order of their energy
(a) 1s, 2s, 3s, 2p

(b) 4s, 3s, 3p, 4d

(c) 5p, 4d, 5d, 4f, 6s

(d) 5f, 6d, 7s, 7p

Based upon the above-given information, solve the following questions given below :

(a) Which among the following orbitals has the lowest energy?

4d, 4f, 5s, 5p

(b) Which among the following orbitals has the highest energy?

5p, 5d, 5f, 6s, 6p

Answer 30. (i) (a) The increasing order of the energy of the given orbital is: 1s >2s >2p> 3s

(b) The increasing order of the energy of the given orbital is: 3s<3p<4s<4d

(c) The increasing order of the energy of the given orbital is: 4d<5p<6s<4f<5d

(d) The increasing order of the energy of the given orbital is: 7s<5f<6d<7p

(ii) (a) For the following orbitals, 5s have the lowest energy.

The (n+l) value of the 5s is the lowest = 5 + 0 = 5.

Other orbitals have (n+l) value of more than 5 –

5p= 5 + 1 = 6 , 4f = 4 + 3 = 7 , 4d = 4 + 2 = 6.

(b) For the following orbitals, 5f has the highest energy as the (n +l) value – 5 + 3 = 8 is the highest.

5d = 5 + 2 = 7 , 5p = 5 + 1= 6 , 6s =6 + 0 = 6 , and 6p =6 + 1 = 7.

Question 31. The bromine atom possesses 35 electrons that contain 6 electrons in 2p orbital, 6 electrons in 3p orbital and 5 electrons in 4p orbital. Which of these electrons experiences the lowest effective nuclear charge?

Answer 31.

The nuclear charge experienced by electrons (which are present in atoms containing multiple electrons) depends on the distance between its orbital and the atom's nucleus. The greater the distance, the lower the effective nuclear charge. Among p-orbitals, 4p orbitals are the farthest from the nucleus of the bromine atom with a (+35) charge. Hence, the electrons that reside in the 4p orbital are the ones to experience the lowest effective nuclear charge. These electrons are also shielded by electrons that are present in the 2p and 3p orbitals along with the s-orbitals.

Question 32. Which orbital would experience the larger effective nuclear charge among the following pairs of orbitals?

(i) 2s and 3s,

(ii) 4d and 4f,

(iii) 3d and 3p

Answer 32.

The nuclear charge can be defined as the net positive charge that acts on an electron in the orbital of an atom with more than 1 electron. It is inversely proportional to the distance between the orbital and the nucleus.

(i) Electrons that reside in the 2s orbital are closer to the nucleus than those in the 3s orbital and, therefore, experience greater nuclear charge.

(ii) 4d orbital is closer to the nucleus than 4f orbital and will experience greater nuclear charge.

(iii) 3p will experience a greater nuclear charge (since it is closer to the nucleus than the 3f orbital).

Question 33. The unpaired electrons in Al and Si are present in the 3p orbital. Which electrons would experience a more effective nuclear charge from the nucleus?

Answer 33.

The nuclear charge can be defined as the net positive charge that acts on an electron in the orbital of an atom with more than 1 electron. The greater the atomic number, the greater the nuclear charge. Silicon holds 14 protons, while aluminum holds only 13. Therefore, the nuclear charge of silicon is greater than that of aluminum, implying that the electrons in the 3p orbital of silicon would experience a more effective nuclear charge than aluminum.

Question 34. Indicate the number of unpaired electrons in:

(a)P

(b)Si

(c)Cr

(d)Fe

(e)Kr

Answer 34.

(a)Phosphorus (P):

The atomic number of phosphorus is 15

Electronic configuration of Phosphorus:

1s 2 2s 2 2p 6 3s 2 3p 3

That can be represented as follows:

From the diagram, it could be observed that phosphorus has three unpaired electrons.

(b) Silicon (Si):

The atomic number of silicon is 14

Electronic configuration of Silicon:

1s 2 2s 2 2p 6 3s 2 3p 2

That can be represented as follows:

From the diagram, it could be observed that silicon has two unpaired electrons.

(c) Chromium (Cr):

The atomic number of Cr is 24

Electronic configuration of Chromium:

$1s^2\ 2s^2\ 2p^6\ 3s^2\ 3p^6\ 4s^1\ 3d^5$

That can be represented as follows:

From the diagram, it could be observed that chromium has six unpaired electrons.

(d) Iron (Fe):

The atomic number of iron is 26

Electronic configuration of Fe:

$1s^2\ 2s^2\ 2p^6\ 3s^2\ 3p^6\ 4s^2\ 3d^6$

That can be represented as follows:

From the diagram, it could be observed that iron has four unpaired electrons.

(e) Krypton (Kr):

The atomic number of Krypton is 36

it is electronic configuration is:

$1s^2\ 2s^2\ 2p^6\ 3s^2\ 3p^6\ 4s^2\ 3d^{10}\ 4p^6$

That can be represented as follows:

From the diagram, it could be observed that Krypton has no unpaired electrons.

Question 35.

(a) How many sub-shells are associated with the n = 4?

(b) How many electrons would be present in the sub-shells having an ms value of –1/2 for n = 4?

Answer 35.

(a) n = 4 (Given)

For a given value of 'n', the values of the 'l' range from 0 to (n – 1).

where the possible values of l are 0, 1, 2, and 3

Hence, four subshells are associated with n = 4, which are s, p, d and f.

(b) Number of orbitals in nth shell = n2

For n = 4

Therefore, the number of orbitals when n = 4 is 16

Each orbital has 1 electron with an ms value of -1/2.

Thus, the total number of electrons present in the subshell having an ms value of (-1/2) is 16.

COMPETITIVE CORNER

NEET

Question 1: What is Bohr's radius of the 4th orbit of He^+ in Å?

    a. 5.23 Å
    b. 3.23 Å

c. 6.23 Å
d. 4.23 Å

Answer: d) 4.23 Å

Explanation: $r_n = a° \times n^2/Z = .53 \text{ Å} \times (4)^2/2 = .53 \text{ Å} \times 16/2$ = **4.23 Å**

Question 2: The energy of radiation emitted when an electron falls from n = 3 to n = 2 level in a hydrogen atom will be: ($R_H = 2.18 \times 10^{-18}$ J atom^{-1})

   a. 0.3×10^{-18} J atom^{-1}
   b. 3×10^{-18} J atom^{-1}
   c. 0.03×10^{-18} J atom^{-1}
   d. 3×10^{-17} J atom^{-1}

Answer: a) 0.3×10^{-18} J atom^{-1}

Explanation:

$\Delta E - E [1/n_1^2 - 1/n_2^2] \times Z^2$

$= 2.18 \times 10^{-18} [1/4 - 1/9] \times 1^2$

$= 2.18 \times 10^{-18} [(9 - 4)/36] \times 1$

$= 2.18 \times 10^{-18} \times 5/36 =$ **0.3×10^{-18} J atom^{-1}**

Question 3: How fast does light travel in vacuum as compared to the velocity of an electron in Bohr's first orbit of hydrogen atom?

    a. 13.7 times
    b. 67 times
    c. 137 times
    d. 97 times

Answer: c) 137 times

Explanation:

$V_n = V^0 \times Z/n$

$\Rightarrow V_n = V^0 \times 1/1$

$V_1 = 2.18 \times 10^6 \text{ ms}^{-1}$

$S = 3 \times 10^8 \text{ ms}^{-1}$

$\therefore (3 \times 10^8) / (2.18 \times 10^6) = 300 / 2.18 =$ **137 times**

Question 4: The first emission lines of the hydrogen atomic spectrum in the Balmer series appear at:

    a. $5R/36 \text{ cm}^{-1}$
    b. $3R/4 \text{ cm}^{-1}$
    c. $7R/144 \text{ cm}^{-1}$
    d. $9R/400 \text{ cm}^{-1}$

Answer: a) $5R/36 \text{ cm}^{-1}$

Explanation: The transition from $n_1 = 2$ to $n_2 = 3$

results in the first emission line in the atomic spectra of hydrogen in the Balmer series.

The wavenumber is given by the expression,

Thus, the first emission lines of the hydrogen atomic spectrum in the Balmer series appear at **5R/36 cm^{-1}**

Question 5: Quantum numbers are different quantities used to describe the electrons present in an atom. Each quantum number signifies a property. Identify the quantum number that determines the orientation of the orbital in space.

    a. Azimuthal quantum number
    b. Spin quantum number
    c. Magnetic quantum number
    d. Principal quantum number

Answer: c) Magnetic quantum number

Explanation: The magnetic quantum number represents the total number of orbitals in a subshell as well as their orientation. The symbol 'm_l' is used to represent it. This value represents the projection of the orbital's angular momentum along a specific axis.

Question 6: Which one of the following about an electron occupying the 1s orbital in a hydrogen atom is incorrect?

(The Bohr radius is represented by a_0)

    a. The total energy of the electron is maximum when it is at a distance a_0 from the nucleus.
    b. The probability density of finding the electron is maximum at the nucleus.
    c. The electron can be found at a distance of $2a_0$ from the nucleus.
    d. The magnitude of the potential energy is double that of its kinetic energy on average.

Answer: a) The total energy of the electron is maximum when it is at a distance a_0 from the nucleus.

Explanation: At a_0 distance from the nucleus, the probability of finding an electron is maximum, and the total energy of the electron is minimum.

Question 7: What is the de Broglie wavelength of the electron accelerated at 400 V, approximately?

    a. 1.00 nm
    b. 0.12 nm
    c. 0.08 nm
    d. 0.06 nm

Answer: d) 0.06 nm

Explanation: Short method:

de Broglie wavelength (λ) = $(12 - 3) / \sqrt{V}$ Å

$\Rightarrow \lambda = (12 - 3) / \sqrt{400}$ Å

$= (12 - 3) / 20$ Å

$= 0.6$ Å

$\therefore \lambda =$ **0.06 nm**

Long method:

Question 8: The uncertainty in the position of an electron is equal to its de Broglie wavelength. The minimum percentage error in its measurement of velocity under this circumstance will be approximate:

    a. 4
    b. 8
    c. 2
    d. 18

Answer: b) 8

Explanation: According to the question, uncertainty in

the position of an electron is equal to its de Broglie wavelength.

$\therefore \Delta x = h / mv$

Now, according to uncertainty principle, $\Delta x * \Delta p = h / 4\pi$

As mentioned in the question, a minimum percent error is required.

So, $m\Delta v * \Delta x = h / 4\pi$

By putting the value of Δx, we get $\Delta v / v = 1 / 4\pi$

$\therefore$ % of error $= \Delta v / v * 100 = 100 / 4\pi = 8$

Question 9: From the following sets of quantum numbers, which state is possible?

(i) $n = 0, l = 0, m_l = 1, m_s = +1/2$

(ii) $n = 2, l = 1, m_l = 0, m_s = -1/2$

(iii) $n = 2, l = 0, m_l = 3, m_s = +1/2$

(iv) $n = 3, l = 1, m_l = 0, m_s = -1/2$

    a. (i), (ii), (iii)
    b. (ii) and (iv) are possible
    c. All are not possible

d. All are possible

Answer: b) (ii) and (iv) are possible

Explanation: (i) Because the minimal value of n can only be 1 and not zero, the set of quantum numbers is not possible.

(ii) This set of quantum numbers is possible.

(iii) If the value of l is zero, then the value of m_l should also start from zero and not 3.

(iv) This set of quantum numbers is also possible.

Question 10: When the 3d orbital is complete, the new electron will enter the:

    a. 4p orbital
    b. 4f orbital
    c. 4s orbital
    d. 4d orbital

Answer: a) 4p orbital

Explanation: The 4p orbital is the next shell to be filled after the 3d orbital, according to the Aufbau principle of electron filling in orbitals.

Question 11: What should be the maximum number of

lines obtained in the spectrum, if the total number of energy levels is five?

   a. 10
   b. 6
   c. 4
   d. 2

Answer: a) 10

Explanation: Given, levels (l) = 5, n_1 = 1, n_2 = 5

∴ Maximum number of lines = ((n2 − n1)(n2 − n1 + 1)) / 2

= (Δn (Δn + 1)) / 2 = (4 x (4 + 1)) / 2 = (4 x 5) / 2 = **10**

Question 12: We can distinguish two electrons occupying the same orbital by:

   a. Principal quantum number
   b. Azimuthal quantum number
   c. Magnetic quantum number
   d. Spin quantum number

Answer: d) spin quantum number

Explanation: The value of a spin quantum number can be either +1/2 or −1/2.

This helps in the differentiation of two electrons

occupying the same orbital.

Question 13: In which quantum level does the electron jump in He$^+$ ion to ground state if it is given an energy corresponding to 99% of the ionisation potential of He$^+$ ion?

    a. 4
    b. 6
    c. 8
    d. 10

Answer: d) 10

Explanation: $E \propto 1/\lambda \propto (1/n_1^2 - 1/n_2^2)$

Here, ionisation potential = 99% (E), $n_1 = 1$ and $n_2 = ?$

$\therefore 99/100 = 1/1 - 1/n_2^2$

$\Rightarrow 1/n_2^2 = 1/100$

$\therefore n_2 = \mathbf{10}$

Question 14: The value of quantum numbers (n, l, m and s, respectively) for the valence electron of Rb (Rubidium) is:

    a. 5, 0, 0, +1/2
    b. 5, 1, 0, +1/2

c. 5, 1, 1, +1/2

d. None of these

Answer: a) 5, 0, 0, +1/2

Explanation:

E.C of Rubidium is (Z=37) = $1s^2\ 2s^2\ 2p^6\ 3s^2\ 3p^6\ 4s^2\ 3d^{10}\ 4p^6\ 5s^1$

The valence electron is in the 5s orbital.

Thus, its 4 quantum numbers are n = 5, l = 0, m = 0, s = +1/2

Question 15: Which statement does not form part of Bohr's model of hydrogen atom?

  a. The energy of the orbit is quantised.
  b. Electrons in orbit nearest to the nucleus have the lowest energy.
  c. Electrons revolve in different orbits around the nucleus.
  d. The position and velocity of electrons in orbit cannot be determined simultaneously.

Answer: d) The position and velocity of electrons in orbit cannot be determined simultaneously.

Explanation: Statement (d) is not included in Bohr's

hydrogen atom model.

An electron in an atom, according to Bohr, is placed at a specific distance from the nucleus and revolves around it at a specific velocity. This contradicts Heisenberg's uncertainty principle, which states that the position and velocity of electrons in orbit cannot be known at the same time.

JEE

1. If the kinetic energy of an electron is increased four times, the wavelength of the de-Broglie wave associated with it would become

(1) Two times

(2) Half

(3) One fourth

(4) Four times

Solution:

The wavelength λ is inversely proportional to the square root of kinetic energy. So if KE is increased 4 times, the wavelength becomes half.

λ ∝ 1/√KE

Hence option (2) is the answer.

2. Calculate the wavelength (in nanometer) associated with a proton moving at $1.0 \times 10^3 \, ms^{-1}$ (Mass of proton = 1.67×10^{-27} kg and h = 6.63×10^{-34} Js)

(1) 2.5 nm

(2) 14.0 nm

(3) 0.032 nm

(4) 0.40 nm

Solution:

Given $m_p = 1.67 \times 10^{-27}$ kg

$h = 6.63 \times 10^{-34}$ Js

$v = 1.0 \times 10^3 \, ms^{-1}$

We know wavelength $\lambda = h/mv$

∴ $\lambda = 6.63 \times 10^{-34} / (1.67 \times 10^{-27} \times 1.0 \times 10^3)$

$= 3.97 \times 10^{-10}$

≈ 0.04nm

Hence option (4) is the answer.

3. The radius of the second Bohr orbit for the hydrogen atom is:

(Planck's constant, $h = 6.262 \times 10^{-34}$ Js: Mass of electron = 9.1091×10^{-31} kg; Charge of electron $e = 1.60210 \times 10^{-19}$ C; permittivity of vacuum $\varepsilon_0 = 8.854185 \times 10^{-12}$ kg^{-1}m^{-3}A^2)

(1) 1.65 A

(2) 4.76 A

(3) 0.529 A

(4) 2.12 A

Solution:

Radius of nth Bohr orbit in H atom = $0.53\, n^2/Z$

For hydrogen Z = 1

Radius of 2nd Bohr orbit in H atom = $0.53 \times 2^2/1 = 2.12$

Hence option (4) is the answer.

4. The energy required to break one mole of Cl–Cl bonds in Cl_2 is 242 kJ mol^{-1}. The longest wavelength of light

capable of breaking a single Cl–Cl bond is

($C = 3 \times 10^8$ m/s and $N_A = 6.02 \times 10^{23}$ mol^{-1})

(1) 494 nm

(2) 594 nm

(3) 640 nm

(4) 700 nm

Solution:

We have B.E = 242KJ/Mol

$E = h_c N_A / \lambda$

$\therefore \lambda = h_c N_A / E$

$= 3 \times 10^8 \times 6.626 \times 10^{-34} \times 6.02 \times 10^{23} / (242 \times 10^3)$

$= 0.494 \times 10^{-3} \times 10^3$

$= 494$ nm

Hence option (1) is the answer.

5. Ionisation energy of He$^+$ is 19.6×10^{-18} J atom^{-1}. The energy of the first stationary state (n = 1) of Li^{2+} is

(1) 8.82×10^{-17} J atom^{-1}

(2) 4.41×10^{-16} J atom^{-1}

(3) -4.41×10^{-17} J atom^{-1}

(4) -2.2×10^{-15} J atom^{-1}

Solution:

Given I.E = 19.6×10^{-18}

I.E $\propto z^2$

(I.E) Li^{2+}/He$^+$ = $(9/4) \times 19.6 \times 10^{-18}$

= -4.41×10^{-17}

Hence the option (3) is the answer.

6. The frequency of light emitted for the transition n = 4 to n = 2 of He+ is equal to the transition in H atom corresponding to which of the following

(1) n = 3 to n = 1

(2) n = 2 to n = 1

(3) n = 3 to n = 2

(4) n = 4 to n = 3

Solution:

$E = 13.6 \times 4[(1/4)-(1/16)]$

$= 10.2$

$E = h\nu$

$\nu = 10.2/h$

$E = 13.6(1)[(1/n_1^2 - 1/n_2^2)]$

$10.2 = 13.6[(1/n_1^2 - 1/n_2^2)]$

$102/136 = (n_2^2 - n_1^2)/n_1^2 n_2^2$

Substitute the given options and find n_1 and n_2

$51/68 = (n_2^2 - n_1^2)/n_1^2 n_2^2$

$0.75 = (4-1)4 = 3/4 = 0.75$

Hence option (2) is the answer.

7. Based on the equation $\Delta E = -2.0 \times 10^{-18}$ J $(1/n_2^2 - 1/n_1^2)$ the wavelength of the light that must be absorbed to excite hydrogen electron from level n = 1 to level n = 2 will be (h = 6.625×10^{-34} Js, C = 3×10^8 ms^{-1})

(1) 2.650×10^{-7} m

(2) 1.325×10^{-7} m

(3) 1.325×10^{-10} m

(4) 5.300×10^{-10} m

Solution:

$\Delta E = -2.0 \times 10^{-18}$ J $(1/n_2^2 - 1/n_1^2)$

$= -2.0 \times 10^{-18}(1/2^2 - 1/1^2)$

$= -2.0 \times 10^{-18}(1/4 - 1/1)$

$= -2.0 \times 10^{-18}(-3/4)$

$= 1.5 \times 10^{-18}$

Also $\Delta E = hc/\lambda$

So $\lambda = hc/\Delta E$

$= 6.625 \times 10^{-34} \times 3 \times 10^8 / 1.5 \times 10^{-18}$

$= 13.25 \times 10^{-8}$

$= 1.325 \times 10^{-7}$ m

Hence option (2) is the answer.

8. The de Broglie wavelength of a car of mass 1000 kg and velocity 36 km/hr is :

$(h = 6.63 \times 10^{-34}$ Js$)$

(1) 6.626×10^{-31} m

(2) 6.626×10^{-34} m

(3) 6.626×10^{-38} m

(4) 6.626×10^{-30} m

Solution:

Given h = 6.63×10^{-34} J/s

m = 1000 kg

v = 36 km/hr = $36 \times 10^3/(60 \times 60)$ m/s = 10 m/s

λ = h/mv

= $6.63 \times 10^{-34}/1000 \times 10$

= 6.63×10^{-38} m

Hence option (3) is the answer.

9. If the binding energy of the electron in a hydrogen atom is 13.6 eV, the energy required to remove the electron from the first excited state of Li^{++} is

(1) 13.6 eV (2) 30.6 eV (3) 122.4 eV (4) 3.4 eV

Solution:

B.E = $13.6 \times Z^2/n^2$, Z is the atomic number and n is the orbital quantum number. For Li^{++}, Z = 3 and n = 2 for the first excited state.

B.E = $13.6 \times 3^2/2^2$

= 30.6 ev

Hence option (2) is the answer.

10. According to Bohr's theory, the angular momentum of an electron in 5th order orbit is

(1) $25\ h/\pi$

(2) $1.0\ h/\pi$

(3) $10\ h/\pi$

(4) $2.5\ h/\pi$

Solution:

n = 5

So angular momentum, = $nh/2\pi$

= $5h/2\pi$

= $2.5\ h/\pi$

Hence option (4) is the answer.

11. The de Broglie wavelength of a tennis ball of mass 60g moving with a velocity of 10m/s is approximately (Planck's constant, h = 6.63×10^{-34} Js)

(1) 10^{-31} m

(2) 10^{-16} m

(3) 10^{-25} m

(4) 10^{-33} m

Solution:

Given m = 60 g

v = 10 m/s

λ = h/mv

= 6.6×10^{-34}/(60×10^{-3}×10) = 10^{-33} m

Hence option (4) is the answer.

12. In a hydrogen atom, if the energy of an electron in the ground state is 13.6 eV, then that in the 2nd excited state is

(1) 1.51 eV

(2) 3.4 eV

(3) 6.04 eV

(4) 13.6 eV

Solution:

The 3rd energy level is the 2nd excited state.

n = 3

$E_n = 13.6/n^2 = 13.6/9 = 1.5$ eV

Hence option (1) is the answer.

13. In the Bohr series of lines of hydrogen spectrum, the third line from the red end corresponds to which one of the following inter-orbit jumps of the electron for Bohr orbits in an atom of hydrogen

(1) 5 → 2

(2) 4 → 1

(3) 2 → 5

(4) 3 → 2

Solution:

The lines falling in the visible spectrum include the Balmer series. So the third line would be $n_1 = 2$ and $n_2 = 5$. Thus the transition is $5 \to 2$

Hence option (1) is the answer.

14. Which of the following sets of quantum numbers is correct for an electron present in 4f orbital?

(1) $n = 4, l = 3, m = +4, s = +\frac{1}{2}$

(2) $n = 3, l = 2, m = -2, s = +\frac{1}{2}$

(3) $n = 4, l = 3, m = +1, s = +\frac{1}{2}$

(4) $n = 4, l = 4, m = -4, s = -\frac{1}{2}$

Solution:

For 4f orbital, $n = 4$ and $l = 3$.

Values of $m = -3, -2, -1, 0, +1, +2, +3$

Hence option (3) is the answer.

15. The number of d-electrons retained in Fe^{2+} (At.no. of Fe = 26) ion is

(1) 4

(2) 5

(3) 6

(4) 3

Solution:

Configuration of Fe^{2+} = $3d^6\ 4s^0$

Hence option (3) is the answer.

16. Which of the following statements in relation to the hydrogen atom is correct?

(1) 3s orbital is lower in energy than 3p orbital

(2) 3p orbital is lower in energy than 3d orbital

(3) 3s and 3p orbitals are of lower energy than 3d orbital

(4) 3s, 3p and 3d orbitals all have the same energy

Solution:

The Auf-bau principle is not applicable for the Hydrogen atom.

Hence option (4) is the answer.

17. Which of the following sets of quantum numbers represents the highest energy of an atom?

(1) n=3, l=2, m=1, s= +½

(2) n=3, l=2, m=1, s= +½

(3) n=4, l=0, m=0, s= +½

(4) n=3, l=0, m=0, s= +½

Solution:

Maximum value of (n +l) represents the highest energy of the orbital.

Hence option (2) is the answer.

18. The outer electron configuration of Gd (Atomic no. 64) is

(1) $4f^4 \, 5d^4 \, 6s^2$

(2) $4f^7 \, 5d^1 \, 6s^2$

(3) $4f^3 \, 5d^5 \, 6s^2$

(4) $4f^8 \, 5d^0 \, 6s^2$

Solution:

Gd shows a half-filled f^7 configuration.

Hence option (2) is the answer.

PERIODIC CLASSIFICATION OF ELEMENTS

Question 1. Consider the following isoelectronic species, Na+, Mg2+, F–and O2–. The correct order of increasing the length of their given radii is _____.

(i) F– < O2- < Mg2+ < Na+

(ii) Mg2+ < Na+< F–< O2-

(iii) O2-< F– < Na+< Mg2+

(iv) O2- < F– < Mg2+ < Na+

Answer 1: Option (ii) is the correct answer.

Question 2. Which among the following is not an actinoid?

(i) Curium (Z = 96)

(ii) Californium (Z = 98)

(iii) Uranium (Z = 92)

(iv) Terbium (Z = 65)

Answer 2: Option (iv) is the correct answer.

Question 3. The order of screening effect of the electrons of s, p, d and f orbitals for the given shell of an atom on their outer shell electrons will:

(i) s > p > d > f

(ii) f > p > s > d

(iii) p < d < s > f

(iv) f > d > p > s

Answer 3: Option (i) is the correct answer.

Question 4. The first ionisation enthalpies of the Na, Mg, Al and Si are in the order as follows:

(i) Na < Mg > Al < Si

(ii) Na > Mg > Al > Si

(iii) Na < Mg < Al < Si

(iv) Na > Mg > Al < Si

Answer 4: Option (i) is the correct answer.

Question 5. The electronic configuration of the given gadolinium (Atomic number 64) is

(i) [Xe] 4f3 5d5 6s2

(ii) [Xe] 4f7 5d2 6s1

(iii) [Xe] 4f7 5d1 6s2

(iv) [Xe] 4f8 5d6 6s2

Answer 5: Option (iii) is the correct answer.

Question 6. The statement which is not correct for the following periodic classification of elements is:

(i) The properties of elements are periodic functions of their given atomic numbers.

(ii) The Non-metallic elements are less in number than the

metallic elements.

(iii) For the given transition elements, the 3d-orbitals fill with the electrons after the 3p-orbitals and before the 4s-orbitals.

(iv) The first ionisation enthalpies of the given elements generally increase with an increase in the atomic number as we go along the period.

Answer 6:

Option (iii) is the correct answer.

Question 7. Among halogens, correct order of the amount of the energy released in the given electron gain (electron gain enthalpy) will:

(i) F < Cl < Br < I

(iv) F > Cl > Br > I

(iii) F < Cl > Br > I

(ii) F < Cl < Br < I

Answer 7: Option (iii) is the correct answer.

Question 8. The period number at the long form of the given periodic table is equal to

(i) the magnetic quantum number of any given element of the period.

(ii) an atomic number of any given element of the period.

(iii) maximum Principal quantum number of any given period element.

(iv) maximum Azimuthal quantum number of any given period element.

Answer 8: Option (iii) is the correct answer.

Question 9. The elements in which the given electrons progressively fill in the 4f-orbitals call

(i) actinoids

(ii) transition elements

(iii) lanthanoids

(iv) halogens

Answer 9: Option (iii) is the correct answer.

Question 10. Which among the following is the correct order of the size of the given species:

(i) I > I– > I+

(ii) I+ > I– > I

(iii) I > I+ > I-

(iv) I– > I > I+

Answer 10: Option (iv) is the correct answer.

Question 11. The formation of the given oxide ion, O^{2-} (g), from the oxygen atom, requires first an exothermic and then an endothermic step as stated below:

O (g) + e– → O–(g) ; thus, Δ HV = – 141 kJ mol–1

O– (g) + e– → O^{2-} (g) ; thus, Δ HV = + 780 kJ mol–1

Thus, the formation process of O^{2-} in the given gas phase is unfavourable even though O^{2-} is isoelectronic with the neon. It is because of this fact that

(i) oxygen is more electronegative.

(ii) addition of the electrons in oxygen results in the larger size of the ion.

(iii) electron repulsion outweighs stability gained by achieving the noble gas configuration.

(iv) O– ion is smaller than a given oxygen atom.

Answer 11:Option (iii) is the correct answer.

Question 12. Electronic configurations of the following four elements, A, B, C and D, have given below :

(A) $1s^2\ 2s^2\ 2p^6$

(B) $1s^2\ 2s^2\ 2p^4$

(C) $1s^2\ 2s^2\ 2p^6\ 3s^1$

(D) $1s^2\ 2s^2\ 2p^5$

Which among the following is the correct order of increasing the tendency to gain the electrons :

(i) A < C < B < D

(ii) D < A < B < C

(iii) D < B < C < A

(iv) A < B < C < D

Answer 12:Option (i) is the correct answer.

Question 13. Which among the following elements can show the covalency greater than the 4?

(i) B

(ii) P

(iii) S

(iv) Be

Answer 13:Option (ii) and (iii) are the correct answers.

Question 14. Those elements which impart colour to the flame on heating in it are the atoms that require low energy for the ionisation (i.e., absorb energy in the visible region of the spectrum). The elements of which among the following groups will impart colour to the flame?

(i) 2

(ii) 13

(iii) 1

(iv) 17

Answer 14:Option (i) and (iii) are the correct answers.

Question 15. Which among the following sequences contains atomic numbers of only representative elements?

(i) 3, 33, 53, 87

(ii) 2, 10, 22, 36

(iii) 7, 17, 25, 37, 48

(iv) 9, 35, 51, 88

Answer 15:Option (i) and (iv) are the correct answers.

Question 16. Which among the following elements will gain one electron more readily in comparison to the other elements of the given group?

(i) S (g)

(ii) Na (g)

(iii) O (g)

(iv) Cl (g)

Answer 16:Option (i) and (iv) are the correct answers.

Question 17. Explain why the electron gain enthalpy of the elemental fluorine is less negative than the elemental chlorine. .

Answer 17: Fluorine has a smaller size as compared to chlorine. As a result, the attraction outside the shell to gain the electrons is less. Moreover, they possess inter-electronic repulsions in the 2p orbitals, resulting in less negative electron gain enthalpy.

Question 18. All the transition elements are d-block elements, but all the d-block elements are not the transition elements. Explain.

Answer 18: The elements having their outermost shell filled with the d electrons are called the d block elements. All the d blocks are not the transition elements as it is important to have an incompletely filled d orbital of the given element like calcium and zinc etc.

Question 19. Identify the group and valency of the following element having atomic number 119. Also, predict the outermost electronic configuration for it and write the general formula of the oxide.

Answer 19: There are 118 elements found in the seven periods of the modern periodic table. Thus, the element with the atomic number 119 will lie in the 8th period of the first group and have the outermost electronic configuration of 8s1. It belongs to group 1 and has the valency one. The formula of the oxide would be M2O.

Question 20. Ionisation enthalpies of a few elements of the second period are given below: Ionisation enthalpy/ kcal mol–1: 520, 899, 801, 1086, 1402, 1314, 1681, 2080.

Here, Match the correct enthalpy by the given elements and complete the graph in the Fig. Also, write symbols of the elements with the given atomic number.

Answer 20:

The ionisation enthalpy of the given elements varies across the period and group. The ionisation enthalpy decreases down the group and increases as we move from left to right in the period.

Question 21. Among the given elements B, Al, C and Si,

(i) which of the following elements has the highest first ionisation enthalpy?

(ii) which element contains the most metallic character? Explain your answer in each case.

Answer 21: (i) Carbon possesses the highest ionisation enthalpy. It increases from left to right along the period and also decreases as we go down the group.

(ii) Aluminium possesses the most metallic character. On moving down, the metallic character increases and decreases across the period from left to right.

Question 22. Write the four characteristic properties for the p-block elements.

Answer 22: 1. They show the variable oxidation states. The trend of the reducing character increases on moving down the group, and oxidising character increases as we move along the period.

They have a higher ionisation enthalpy than the s-block elements.
They usually form the given covalent compounds.
Both metals and non-metals can be found in this group, but the non-metals are slightly more in number.

Question 23. Illustrate by taking examples of the following transition and non-transition elements for which the oxidation states of the given elements are mostly based on the electronic configuration.

Answer 23: Ti possesses an atomic number of 22 and electronic configuration [Ar]3d24s2 and has three oxidation states at +2,+3 and +4 in the various compounds like TiO29(+4), Ti2O3(+3) and TiO(+2). The non-transition elements like the p-block elements have variable oxidation states like the phosphorus. It has -3,+3 and +5.

Question 24. Nitrogen possesses the positive electron gain enthalpy, whereas the oxygen possesses the negative. But oxygen has lower ionisation enthalpy than nitrogen. Explain.

Answer 24: The ionisation enthalpy of oxygen is lower than that of nitrogen as when we remove one electron from the oxygen then it easily donates it to attain half-filled stability; however, in the case of nitrogen, it is difficult to remove one electron because it already has half-filled stability and it will become unstable after that.

Question 25. The first member of each group of the representative elements, s and p-block elements, shows abnormal behaviour. Illustrate with the help of two examples.

Answer 25: Lithium and beryllium are the examples. Li is the first group element. It has different properties as well as forms the covalent compounds as well as nitrides. Beryllium is the first element of the second group. It has various anomalies like it forms the covalent compound with coordination number four, unlike other elements with the coordination number 6.

Question 26. What is the basic theme for the organisation in the modern periodic table?

Answer 26. The basic theme of organisation in the modern periodic table is to classify the given elements in periods and groups according to its properties. Hence, the course of action makes the study of elements and their compounds simple and systematic. In the periodic table, elements with similar properties are placed in the same group.

Question 27. How does the metallic and non-metallic character vary as we move from left to right in the period?

Answer 27:

Metallic character decreases as we move along left to right across the period, and non-metallic character increases as you can find an increase in the ionisation enthalpy and electron gain enthalpy along the period.

Question 28. The radius of the Na+ cation is less than that of the Na atom. Give reasons for your answer.

Answer 28:

Sodium atoms can lose one electron to form sodium cation. After the formation of the cation, the effective nuclear charge on the ion increases on the left electrons, resulting in the decrease of the radius.

Question 29. Among alkali metals which element can be least electronegative and why?

Answer 29:

Caesium is the least electronegative alkali metal because electronegativity decreases as we move from top to bottom due to increase in size.. Caesium is the group 1 element and lies down the group because it has the largest size due to the decrease in the effective nuclear charge.

Question 30. What do you understand about the term exothermic reaction and endothermic reaction? Give one example of each type.

Answer 30: Exothermic reaction- The reaction where heat evolved is called an exothermic reaction.

for example,

CaO + CO2 → CaCO3 ΔH=-178 kJmol-1

Endothermic reaction- The reaction where heat is absorbed is called an endothermic reaction.

for example,

2NH3 → 3N2+ H2 ΔH=918kJmol-1

Question 31. Arrange the following elements N, P, O and S in the order of-

(i) increasing first ionisation enthalpy.

(ii) increasing non-metallic character.

Give an appropriate reason for the arrangement assigned.

Answer 31:

(i) S< P< O< N is the accurate, increasing order of the first ionisation enthalpy.

On going down the group, the ionisation enthalpy decreases, and as we move along the period, then it increases; however, in the case of oxygen and nitrogen, because of the half-filled stability of 2p orbitals of nitrogen, it has the higher ionisation enthalpy than oxygen.

(ii) P<S<N<O is the accurate, increasing order of non-metallic character.

Moving down the group, we will see non-metallic character decrease as the effective nuclear charge on the outermost shell decreases, which helps to gain an electron. The effective nuclear charge increases moving along the period, increasing the non-metallic character.

Question 32. p-Block elements can form acidic, basic and amphoteric oxides. Explain each given property by giving two examples and write the reactions of these oxides with water.

Answer 32:

ACIDIC OXIDES

SO_2 and B_2O_3 are the acidic oxides in the p block elements.

The required reaction of SO_2 with water

$SO_2 + H_2O \rightarrow H_2SO_3$

The required reaction of B_2O_3 with water

$B_2O_3 + 3 H_2O \rightarrow 2H_3BO_3$

Acidic oxides are those oxides that can form acids after reacting with water.

BASIC OXIDES

Cao, BaO, and Ti_2O form the basic oxides

The required reaction of Ti_2O with water.

$BaO + H_2O \rightarrow Ba(OH)_2$

REACTION OF CaO WITH WATER

$Cao + H_2O \rightarrow Ca(OH)_2$

Basic oxides are those oxides that can form bases after reacting with the water.

AMPHOTERIC OXIDES

Zinc oxides, as well as aluminium oxides, are the two given amphoteric oxides.

Reaction of ZnO with the water:

$ZnO + 2H_2O + 2NaOH \rightarrow Na_3Zn[OH]_4 + H_2$

$ZnO + 2HCl \rightarrow ZnCl_2 + H_2O$

Reaction of Al_2O_3 with the water:

Al2O3 (s) + 6 NaOH(aq) + 3H2O(l) → Na3 [Al(OH)6] (aq)

Al2O3 (s) + 6HCl(aq) + 9H2O(l) → 2[Al(H2O)]3+(aq) + 6Cl−

Question 33. Explain the deviation in the ionisation enthalpy of some given elements from the general trend using Fig.

Answer 33:

The ionisation enthalpy of given elements varies across periods and groups. The ionisation enthalpy decreases down the group and increases as we go on from left to right in the period.

Effective nuclear charge for the outermost electrons.
Electron-electron repulsion force.
Stability of the element because of half-filled and filled orbitals are some of the parameters affected.

Question 34. Explain the following:

(a) Electronegativity of the given elements increases as we move from left to right in the periodic table.

(b) Ionisation enthalpy decreases in a group from top to bottom?

Answer 34:

(a) As we move left to right in the period, the size of the atoms decreases because of the increase in the effective nuclear charges on the outermost electron. As a result, the electronegativity of the elements increases as we move along left to right in the periodic table.

(b) we move down the group, then the atomic size increases, which results by the increase in the distance of the electrons in the outermost shell. As a result, the effective nuclear charge decreases. That results in the decrease of the ionisation enthalpy.

Question 35. Which important properties did Mendeleev use to classify the elements in his periodic table, and did he stick to that?

Answer 35. Mendeleev organised the components in his periodic table according to the order of their atomic weight. He organised the components in groups and periods according to the increment order of atomic weight. He placed the elements with equal properties in the same group.

So, he did not stick to this arrangement for long. He discovered that if the elements were organised according to their increasing atomic weights, thus some of the elements didn't fit in with his classification scheme.

Thus, he ignored the order of atomic weights in some cases. For example, the atomic mass of the iodine is lower than the atomic mass of the tellurium.

Still, Mendeleev set tellurium (in the Group 6) ahead of iodine (in Group 7) along with fluorine, chlorine, and bromine because of similarities in properties.

Question 36. What is the basic difference in the required approach between Mendeleev's Periodic Law and the Modern Periodic Law?

Answer 36.

Mendeleev's Periodic law   Modern Periodic Law
Mendeleev's Periodic Law states that elements' physical and chemical properties are periodic functions of their atomic weights.   Modern Periodic Law states that elements' physical and chemical properties are periodic functions of their atomic numbers.

Question 37. Based on quantum numbers, justify that the sixth period of the periodic table should have 32 elements.

Answer 37. In the periodic table containing elements, a period shows the value of a principal quantum number (n) for the outermost shells. Every period starts with filling the principal quantum number (n). The value for n for the 6th period is equal to 6. then, for n = 6, the azimuthal quantum number (l) could have "0, 1, 2, 3, 4" values.

According to Aufbau's principle, electrons have been added to the various orbitals in order of their increasing energies. Here, the 6d subshell has much higher energy than the energy of the 7s subshell.

In the sixth period, the electrons could fill in only 6s, 4f, 5d, and 6p subshells. 6s has only one orbital, 4f has seven orbitals, 5d has five orbitals, also 6p has three orbitals. Therefore, there are a sum of 16 (1 + 7 + 5 + 3 = 16) orbitals available. As per Pauli's exclusion principle, each orbitals can accommodate a maximum of 2 electrons.

Hence, sixteen orbitals can accommodate a maximum of 32 electrons.

So, the 6th period of the periodic table should have 32 elements.

Question 38. In terms of the period and group, where would you locate the element with Z =114?

Answer 38. Elements of the atomic numbers from Z = 87 to the Z = 114 are present in the seventh period of the periodic table. Hence, the element with Z = 114 is present in the seventh period for the periodic table.

At the seventh period, first 2 elements with Z = 87 and Z= 88 are s-block elements and the next 14 elements except Z = 89 i.e., those with Z = 90 to Z = 103 are f – block elements, as well as next 10 elements with Z = 89 as well as Z = 104 to Z = 112 are d-block elements as well as the elements with Z = 113 to Z = 118 are the p-block elements. Thus, the element with Z = 114 is the second p-block element in the seventh period. So, the element with Z = 114 is present in the periodic table's seventh period and fourteenth group.

Therefore,

Period = 7th, Group = 14 and Block = p-block.

Question 39. In the modern periodic table, elements are arranged in order of their increasing atomic number related to the electronic configuration. Depending upon the type of the orbitals that receive the last electron, the periodic table elements are categorised into four blocks, viz, s, p, d and f. The modern periodic table has 7 periods and 18 groups. Each period starts with the filling of the new energy shell. By the Aufbau principle, the seven periods (1 to 7) contain 2, 8, 8, 18, 18, 32 and 32 elements, respectively. The seventh period is still incomplete. To avoid today's periodic table being too long, the two series of the f-block elements, called lanthanoids and actinoids, are placed at the bottom of the main body of today's periodic table.

(a) The element having the atomic number 57 belongs to

(i) s-block

(ii) p-block

(iii) d-block

(iv) f-block

(b) The outermost electronic configuration represents the last element of the existing p-block in the 6th period.

(i) $4f^{14} 5d^{10} 6s^2 6p^4$

(ii) $5f^{14} 6d^{10} 7s^2 7p^0$

(iii) $4f^{14} 5d^{10} 6s^2 6p^6$

(iv) $7s^2 7p^6$

(c) Which of the following elements whose atomic numbers given below cannot be accommodated in the present set-up of the long form of the periodic table?

(i) 102

(ii) 118

(iii) 126

(iv) 107

(d) The electronic configuration of the given element that is just above the element with the atomic number 43 in the same group will _____.

(i) 1s2 2s2 2p6 3s2 3p6 3d5 4s2

(ii) 1s2 2s2 2p6 3s2 3p6 3d6 4s2

(iii) 1s2 2s2 2p6 3s2 3p6 3d7 4s2

(iv) 1s2 2s2 2p6 3s2 3p6 3d5 4s3 4p6

(e) The elements having atomic numbers 35, 53 and 85 will all _____.

(i) light metals

(ii) halogens

(iii) heavy metals

(iv) noble gases

Answer 39:

(a) d-block,

(b) 4f 14 5d10 6s2 6p6

(c) 126

(d) 1s2 2s2 2p6 3s2 3p6 3d5 4s2

(e) halogens

Question 40. What is the atomic number of elements keeping in mind both the cases given below;

Element is in the 3rd period of the periodic table.
Element is in the 17th group of the periodic table.
Answer 40. In the third period, we provided that element. The element's highest principal quantum number (n) into which the last electron enters is called the period number. As a result, n=3 for the 3rd period. In addition, the element could be found in the seventeenth group. The elements for the seventeenth group have the following general configuration: ns2np5.

As a result, the needed element's overall electronics configuration is 3s23p5 (because n=3).

The element's complete electrical configuration will now be: 1s22s22p63s23p5.

then add up the total number of electrons in the element's ground state: 1+2+6+2+5=17 electrons.

The element's atomic number equals its ground state's total number of electrons.

As determined above, the element has 17 electrons in its ground state; its atomic number is 17. Chlorine is an atomic number 17 element (Cl).

Question 41. Discuss those factors affecting electron gain enthalpy as well as the trend in its variation in the periodic table.

Answer 41. factors affecting electron gain enthalpy as well as the trend in its variation in the periodic table are:

1)ATOMIC SIZEAs, we go down in the group, the electron gain enthalpy decreases as the distance of the nucleus from outermost shell increases, which decreases its tendency to gain electrons and electron gain enthalpy becomes less negative.

2) EFFECTIVE NUCLEAR CHARGE As we go from the left to right in a period, the effective nuclear charge increases, as well as when we move down the group, it decreases, that results in the attraction of the electrons from the outermost shell

3) ELECTRONIC CONFIGURATION The tendency to gain electrons depends upon the stability of the element. Elements having complete or half-filled stable orbitals have a low tendency to gain electrons; thus, they have very low electron gain enthalpy. TRENDS Across a period, electron gain enthalpy becomes more negative. Down the group, the electron gain enthalpy becomes less negative.

Question 42. Define the ionisation enthalpy. Discuss the factors affecting ionisation enthalpy of elements and their trends in the periodic table.

Answer 42. Ionisation enthalpy is the energy required by the isolated and gaseous atom in its ground state to remove an electron. The effective nuclear charge is due to the screening effect; inner core electrons shield the valence electrons. eThe effective nuclear charge is less than the actual charge on the atom. Penetrated orbital: It is difficult to remove an electron from the orbitals closer to the nucleus and penetrate towards the nucleus. The order of the penetration has been given by s>p>d>f Stability of the orbitals: Half-filled and filled orbital have a high ionisation enthalpy as well as they don't want to lose their stability. Across a period, ionisation enthalpy increases along with the period. Down the group, the ionisation enthalpy decreases.

Answer 42. Ionisation enthalpy is the energy required by the isolated and gaseous atom in its ground state to remove an electron. The effective nuclear charge is due to the screening effect; inner core electrons shield the valence electrons. eThe effective nuclear charge is less than the actual charge on the atom. Penetrated orbital: It is difficult to remove an electron from the orbitals closer to the nucleus and penetrate towards the nucleus. The order of the penetration has been given by s>p>d>f Stability of the orbitals: Half-filled and filled orbital have a high ionisation enthalpy as well as they don't want to lose their stability. Across a period, ionisation enthalpy increases along with the period. Down the group, the ionisation enthalpy decreases.

Question 43. Assertion (A): ionisation enthalpy increases from the left to right in a period.

Reason (R): When the successive electrons added to the orbital in the same principal quantum level, the shielding effect of the inner core of electrons does not increase much to compensate for increased attraction of the electron to the nucleus.

(i) The assertion is the correct statement, and the reason is the wrong statement.

(ii) Assertion and reason both were correct statements, and the reason is the correct explanation of assertion.

(iii) Assertion and reason both were wrong statements.

(iv) The assertion is the wrong statement, and the reason is the correct statement.

Answer 43. (ii) Assertion and reason both were correct statements, and the reason is the correct explanation of assertion.

Question 44. Assertion (A): Boron has a smaller first ionisation enthalpy than beryllium.

Reason (R): The penetration of the 2s electron to the nucleus is more than the 2p electron; therefore, the 2p electron is more shielded by the inner core of the electron than 2s electrons.

(i) Assertion and reason both were correct statements, but the reason is not the correct explanation for the assertion.

(ii) The assertion was a correct statement, but the reason was the wrong statement.

(iii) Assertion and reason both were correct statements, and the reason is the correct explanation for the assertion.

(iv) Assertion and reason both were wrong statements.

Answer 44. (iii) Assertion and reason both were correct statements, and the reason is the correct explanation for the assertion.

Question 45. Assertion (A): Electron gain enthalpy becomes less negative as we go down a group.

Reason (R): The size of the atom increases on going down the group, and the added electron would be farther from the nucleus.

(i) Assertion and reason both were correct statements, but the reason is not the correct explanation for the assertion.

(ii) Assertion and reason both were correct statements, and the reason is the correct explanation for the assertion.

(iii) Assertion and reason both were wrong statements.

(iv) The assertion was the wrong statement, but the reason was the correct statement.

Answer 45. (iv) The assertion was the wrong statement, but the reason was the correct statement.

Question 46. How will you explain that the first ionisation enthalpy of sodium is quite lower than that of magnesium; however, its second ionisation enthalpy is quite higher than that of magnesium?

Answer 46: Sodium attains the stable configuration if it loses on an electron from the outermost shell. That is why its first ionisation enthalpy is less than the magnesium. However, in the case of the second ionisation, magnesium has one electron in its outermost shell to attain stability; it loses the electron easily compared to sodium, which is already stable.

COMPETITIVE CORNER

NEET

1. Find the successive elements of the periodic table with ionisation energies, 2372, 520 and 890 kJ per mol respectively

(a) Li, Be, B

(b) H, He, Li

(c) B, C, N

(d) He, Li, Be

Answer: (d)

2. In the modern periodic table, the number of period of the element is the same as

(a) principal quantum number

(b) atomic number

(c) azimuthal quantum number

(d) atomic mass

Answer: (a)

3. The correct order for the size of I, I^+, I^- is

(a) $I > I^- > I^+$

(b) $I > I^+ > I^-$

(c) $I^- > I > I^+$

(d) $I^+ > I^- > I$

Answer: (c)

4. For the same value of n, the penetration power of orbital follows the order

(a) $s = p = d = f$

(b) $p > s > d > f$

(c) $f < d < p < s$

(d) $s < p < d < f$

Answer: (c)

5. Which of the reactions will need the maximum

amount of energy?

(a) Na → Na$^+$ + e$^-$

(b) Ca$^+$ → Ca^{++} + e$^-$

(c) K$^+$ → K^{++} + e$^-$

(d) C^{2+} → C^{3+} + e$^-$

Answer: (c)

6. Which of the following statements is incorrect?

(a) I.E.$_1$ of O is lower than that of N but I.E.$_2$ O is higher than that of N

(b) The enthalpy of N to gain an electron is almost zero but of P is 74.3 kJ mol^{-1}

(c) isoelectronic ions belong to the same period

(d) The covalent radius of iodine is less than its Van der Waal's radius

Answer: (c)

7. The correct order of electronegativity is

(a) Cl > F > O > Br

(b) F > O > Cl > Br

(c) F > Cl > Br > O

(d) O > F > Cl > Br

Answer: (b)

8. Two different beakers contain M_1-O-H, and M_2-O-H solutions separately. Find the nature of the two solutions if the electronegativity of M_1 = 3.4, M_2 = 1.2, O = 3.5, H = 2.1

(a) acidic, acidic

(b) basic, acidic

(c) basic, basic

(d) acidic, basic

Answer: (d)

9. Which one is the most acidic among these?

(a) MgO

(b) CaO

(c) Al_2O_3

(d) Na_2O

Answer: (c)

10. Which one will have the highest 2nd ionisation energy?

(a) $1s^2 2s^2 2p^6 3s^1$

(b) $1s^2 2s^2 2p^4$

(c) $1s^2 2s^2 2p^6$

(d) $1s^2 2s^2 2p^6 3s^2$

Answer: (a)

JEE

1. The bond dissociation energy of B–F in BF_3 is 646 kJ mol^{-1} whereas that of C–F in CF_4 is 515 kJ mol^{-1}. The correct reason for higher B–F bond dissociation energy as compared to that of C–F is

(1) Significant $p\pi - p\pi$ interaction between B and F in BF_3 whereas there is no possibility of such interaction between C and F in CF_4.

(2) Lower degree of $p\pi - p\pi$ interaction between B and F

in BF$_3$ than that between C and F in CF$_4$

(3) Smaller size of B-atom as compared to that of C-atom

(4) Stronger bond between B and F in BF$_3$ as compared to that between C and F in CF$_4$.

Solution:

Because of pπ – pπ back bonding in BF$_3$ molecule, all B-F bonds have partial double bond character.

Hence option (1) is the answer.

2. Among the following species which two have a trigonal bipyramidal shape?

(1) NI$_3$ (2) I$_3^-$ (3) SO$_3^{2-}$ (4) NO$_3^-$

(1) II and III

(2) III and IV

(3) I and IV

(4) I and III

Solution:

Let us find the hybridization (H) and shape of given species.

(1) For NI_3, $H = ½ (5+3) = 8/2 = 4 \rightarrow sp^3$ hybridized state. It is trigonal pyramidal in shape.

(2) For I_3^-, $H = ½ (7+2+1) = 10/2 = 5 \rightarrow sp^3d$ hybridized state. It is linear in shape.

(3) For SO_3^{2-}, $H = ½ (6+2) = 8/2 = 4 \rightarrow sp^3$ hybridized state. It is trigonal pyramidal in shape.

(4) For NO_3^-, $H = ½ (5+1) = 6/2 = 3 \rightarrow sp^2$ hybridized state. It is trigonal planar in shape.

Hence option (4) is the answer.

3. Using MO theory, predict which of the following species has the shortest bond length?

(1) O_2^-

(2) O_2^{2-}

(3) O_2^{2+}

(4) O_2^+

Solution:

Chemical species	O_2^-	O_2^{2-}	O_2^{2+}	O_2^+
Bond order	1.5	1	3	2.5

Therefore bond length order $O_2^{2-} > O_2^- > O_2^+ > O_2^{2+}$

Hence option (3) is the answer.

4. Among the following, the species having the smallest bond is :

(1) NO

(2) NO^+

(3) O_2

(4) NO^-

Solution:

The bond order of given molecules are:

NO = 2.5, NO^+ = 3, O_2 = 2, NO^- = 2

Larger the bond order, the smaller the bond length.

NO^+ has the largest bond order 3.

Therefore, it will have the smallest bond.

Hence option (2) is the answer.

5. The hybridisation of orbitals of N atom in NO_3^-, NO_2^+, NH_4^+ are respectively:

(1) sp^2, sp^3, sp

(2) sp, sp^3, sp^2

(3) sp, sp^2, sp^3

(4) sp^2, sp, sp^3

Solution:

In NO_3, the central N atom has 3 bonding domains and zero lone pairs of electrons.

In NO_2, the central N atom has 2 bonding domains and zero lone pairs of electrons.

In NH_4, the central N atom has 4 bonding domains and zero lone pairs of electrons.

The Hybridization of N atom in NO_3^-, NO_2^+, NH_4^+ are sp^2, sp, sp^3 respectively.

Hence option (4) is the answer.

6. Based on lattice energy and other considerations, which one of the following alkali metal chlorides is

expected to have the highest melting point?

(1) RbCl

(2) LiCl

(3) KCl

(4) NaCl

Solution:

NaCl has the highest melting point.

Hence option (4) is the answer.

7. The structure of IF_7 is :

(1) octahedral

(2) pentagonal bipyramid

(3) square pyramid

(4) trigonal bipyramid

Solution:

For IF_7, hybridisation – sp^3d^3. The shape is pentagonal bipyramidal.

Hence option (2) is the answer.

8. Which of the following has the square planar structure:

(1) NH_4^+

(2) CCl_4

(3) XeF_4

(4) BF_4^-

Solution:

Hybridization of XeF_4 sp^3d^2

It has a square planar shape.

Hence option (3) is the answer.

9. Among the following the maximum covalent character is shown by the compound:

(1) $AlCl_3$

(2) $MgCl_2$

(3) $FeCl_2$

(4) $SnCl_2$

Solution:

Al^{+3} is having the highest polarizing power than other compounds having greater covalent character.

Hence option (1) is the answer.

10. The compound of Xenon with zero dipole moment is :

(1) XeO_3

(2) XeO_2

(3) XeF_4

(4) $XeOF_4$

Solution:

XeF_4 has dipole moment zero.

Hence option (3) is the answer.

11. Which of the following has a maximum number of lone pairs associated with Xe?

(1) XeO_3

(2) XeF_4

(3) XeF_6

(4) XeF_2

Solution:

XeO_3 has 1 lone pair of electrons. XeF_4 has 2 lone pairs of electrons. XeF_6 has 1 lone pair of electrons. XeF_2 has 3 lone pairs of electrons. XeF_2 has a maximum number of lone pairs of electrons.

Hence option (4) is the answer.

12. Among the following the molecule with the lowest dipole moment is :

(1) $CHCl_3$

(2) CH_2Cl_2

(3) CCl_4

(4) CH_3Cl

Solution:

The order of the dipole moment is $CCl_4 < CHCl_3 < CH_2Cl_2 < CH_3Cl$. So CCl_4 has the lowest dipole moment.

Hence option (3) is the answer.

13. The number of types of bonds between two carbon

atoms in calcium carbide is

(1) One sigma, two pi

(2) One sigma, one pi

(3) Two sigma, one pi

(4) Two sigma, two pi

Solution:

$CaC_2 \rightarrow Ca^{+2} + C_2^{2-}$

$^-C \equiv C^-$

Number of sigma bond is 1 and number of pi bond is 2.

Hence option (1) is the answer.

14. The formation of molecular complex BF_3 – NH_3 results in a change in the hybridisation of boron

(1) From sp^3 to sp^3d

(2) From sp^2 to dsp^2

(3) From sp^3 to sp^2

(4) From sp^2 to sp^3

Solution:

In BF_3, Boron atom has 3 bond pairs of electrons and 0 lone pairs of electrons. It is sp^2 hybridized. In $F_3B \leftarrow NH_3$, Boron atom has 4 bond pairs of electrons and 0 lone pairs of electrons. It is sp^3 hybridized. So the formation of molecular complex results in a change in the hybridization of boron from sp^2 to sp^3.

Hence option (4) is the answer.

15. The molecule having the smallest bond angle is :

(1) PCl_3

(2) NCl_3

(3) $AsCl_3$

(4) $SbCl_3$

Solution:

Bond angle order $NCl_3 > PCl_3 > AsCl_3 > SbCl_3$.

Hence option (4) is the answer.

16. In which of the following pairs the two species are not isostructural?

(1) AlF_6^{3-} and SF_6

(2) CO_3^{2-} and NO_3^-

(3) PCl_4^+ and $SiCl_4$

(4) PF_5 and BrF_5

Solution:

PF_5 has a trigonal bipyramidal shape. BrF_5 has a square pyramidal shape.

Hence option (4) is the answer.

17. Which one of the following molecules is expected to exhibit diamagnetic behaviour?

(1) C_2

(2) N_2

(3) O_2

(4) S_2

Solution:

C_2 and N_2 have no unpaired electrons. So they exhibit diamagnetic behaviour.

18. Which of the following is the wrong statement?

(1) ONCl and ONO⁻ are not isoelectronic

(2) O_3 molecule is bent

(3) Ozone is violet-black in solid-state

(4) Ozone is diamagnetic gas

Solution:

In the given options all are correct statements.

19. Stability of the species Li_2, Li_2^- and Li_2^+ increases in the order of :

(1) $Li_2 < Li_2^+ < Li_2^-$

(2) $Li_2^- < Li_2^+ < Li_2$

(3) $Li_2 < Li_2 < Li_2^+$

(4) $Li_2^- < Li_2 < Li_2^+$

Solution:

The bond order of Li_2 is 1. The bond order of Li_2^+ is 0.5. The bond order of Li_2^- is 0.5. Stability will depend on the bond order. Li_2^+ is more stable than Li_2^- because the higher interelectronic repulsion in Li_2^- makes it the least stable. So the order is $Li_2 > Li_2^+ > Li_2^-$.

Hence option (2) is the answer.

20. In which of the following pairs of molecules/ions, both species are not likely to exist?

(1) H_2^+, He_2^{2-}

(2) H_2^-, He_2^{2-}

(3) H_2^{2+}, He_2

(4) H_2^-, He_2^{2+}

Solution:

The bond order of H_2^{2+} and He_2 is zero. So these molecules do not exist.

Hence option (3) is the answer.

21. Bond distance in HF is 9.17×10^{-11} m. Dipole moment of HF is 6.104×10^{-30} Cm. The per cent ionic character in HF will be : (electron charge = 1.60×10^{-19} C)

(1) 61.0%

(2) 38.0%

(3) 35.5%

(4) 41.5%

Solution:

Given Bond distance = 9.17×10^{-11} m.

Dipole moment = 6.104×10^{-30} Cm

% iconic character = $6.104 \times 10^{-30} \times 100 / (1.60 \times 10^{-19} \times 9.17 \times 10^{-11})$

= 41.5%

Hence option (4) is the answer.

22. In which of the following ionization processes the bond energy has increased and also the magnetic behaviour has changed from paramagnetic to diamagnetic?

(1) $NO \rightarrow NO^+$

(2) $O_2 \rightarrow O_2^+$

(3) $N_2 \rightarrow N_2^+$

(4) $C_2 \rightarrow C_2^+$

Solution:

During the ionisation of $NO \rightarrow NO^+$, the bond order changes from 2.5 to 3. Also magnetic character changes from paramagnetic to diamagnetic.

During the ionisation of $O_2 \rightarrow O_2^+$, the bond order increases from 2 to 2.5 and the magnetic character changes from paramagnetic to diamagnetic.

During the ionisation of $N_2 \rightarrow N_2^+$, the bond order decreases from 3 to 2.5 and the magnetic behaviour changes from diamagnetic to paramagnetic.

During the ionisation of $C_2 \rightarrow C_2^+$, the bond order decreases from 2 to 1.5 and the magnetic behaviour changes from diamagnetic to paramagnetic.

Hence option (1) is the answer.

23. Which one of the following molecules is paramagnetic?

(1) NO

(2) O_3

(3) N_2

(4) CO

Solution:

NO has an unpaired electron. So it is paramagnetic in nature.

Hence option (1) is the answer.

24. The catenation tendency of C, Si and Ge is in the order Ge < Si < C. The bond energies (in kJ mol^{-1} of C—C, Si—Si and Ge—Ge bonds are respectively :

(1) 348, 260, 297

(2) 348, 297, 260

(3) 297, 348, 260

(4) 260, 297, 348

Solution:

Bond energy order is C – C > Si – Si > Ge – Ge.

Hence option (2) is the answer.

25. Oxidation state of sulphur in anions SO_3^{2-}, $S_2O_4^{2-}$ and $S_2O_6^{2-}$ increases in the orders

(1) $S_2O_6^{2-} < S_2O_4^{2-} < SO_3^{2-}$

(2) $SO_3^{2-} < S_2O_4^{2-} < S_2O_6^{2-}$

(3) $S_2O_4^{2-} < SO_3^{2-} < S_2O_6^{2-}$

(4) $S_2O_4^{2-} < S_2O_6^{2-} < SO_3^{2-}$

Solution:

The oxidation state of sulphur in SO_3^{2-} is +4. The Oxidation state of sulphur in $S_2O_4^{2-}$ is +3 and in $S_2O_6^{2-}$ is +5. So the order is $S_2O_4^{2-} < SO_3^{2-} < S_2O_6^{2-}$

Hence option (3) is the answer.

26. In which of the following species is the underlined carbon having sp^3 hybridisation?

(1) CH_3COOH

(2) CH_3CH_2OH

(3) CH_3COCH_3

(4) $CH_2=CH-CH_3$

Solution:

Only in CH_3CH_2OH, carbon has sp^3 hybridisation.

In other molecules, the carbon atom has multiple bonds,

Hence option (2) is the answer.

27. In which of the following sets, all the given species are isostructural?

(1) BF_3, NF_3, PF_3, AlF_3

(2) PCl_3, $AlCl_3$, BCl_3, $SbCl_3$

(3) BF_4^-, CCl_4, NH_4^+, PCl_4^+

(4) CO_2, NO_2, ClO_2, SiO_2

Solution:

BF_4^-, CCl_4, NH_4^+, PCl_4^+ are tetrahedral.

Hence option (3) is the answer.

28. In XeF_2, XeF_4, XeF_6 the number of lone pairs of Xe are respectively

(1) 2, 3, 1

(2) 1, 2, 3

(3) 4, 1, 2

(4) 3, 2, 1

Solution:

XeF_2 has 3 lone pairs of electrons. XeF_4 has 2 lone pairs of electrons. XeF_6 has 1 lone pair of electrons.

Hence option (4) is the answer.

29. Which of the following statements is true?

(1) HF is less polar than HBr

(2) absolutely pure water does not contain any ions

(3) chemical bond formation take place when forces of attraction overcome the forces of repulsion

(4) in covalency transference of electron takes place

Solution:

Chemical bond formation takes place when forces of attraction overcome the forces of repulsion.

Hence option (3) is the answer.

30. Which one of the following pairs of molecules will have permanent dipole moments for both members?

(1) NO_2 and CO_2

(2) NO_2 and O_3

(3) SiF_4 and CO_2

(4) SiF_4 and NO_2

Solution:

NO_2 and O_3 have angular shapes. So they will have a net dipole moment.

Hence option (2) is the answer.

31. The states of hybridization of boron and oxygen atoms in boric acid (H_3BO_3) are respectively

(1) sp^2 and sp^2 (2) sp^3 and sp^3

(3) sp^3 and sp^2 (4) sp^2 and sp^3

Solution:

Hybridization of B is sp^2 and O is sp^3

Hence option (4) is the answer.

32. The maximum number of 90° angles between bond pair of electrons is observed in

(1) dsp^3 hybridization

(2) sp^3d^2 hybridization

(3) dsp^2 hybridization

(4) sp^3d hybridization

Solution:

sp³d² hybridisation has an octahedral configuration. All the bond angles are 90° in the structure.

Hence option (2) is the answer.

33. Which of the following are arranged in an increasing order of their bond strengths?

(1) $O_2^- < O_2 < O_2^+ < O_2^{2-}$

(2) $O_2^{2-} < O_2^- < O_2 < O_2^+$

(3) $O_2^- < O_2^{2-} < O_2 < O_2^+$

(4) $O_2^+ < O_2 < O_2^- < O_2^{2-}$

Solution:

Higher the bond order, stronger the bonds. The increasing order is $O_2^{2-} < O_2^- < O_2 < O_2^+$.

Hence option (2) is the answer.

34. Bond order and magnetic nature of CN⁻ are respectively

(1) 3, diamagnetic

(2) 2.5, paramagnetic

(3) 3, paramagnetic

(4) 2.5, diamagnetic

Solution:

Bond order = ½ [$n_b - n_a$]

= ½ [10-4]

= ½ (6)

= 3

It does not have unpaired electrons. So, it is diamagnetic.

Hence option (1) is the answer.

35. The bond order in NO is 2.5 while that in NO^+ is 3. Which of the following statements is true for these two species?

(1) Bond length in NO^+ is greater than in NO

(2) Bond length is unpredictable

(3) Bond length in NO^+ in equal to that in NO

(4) Bond length in NO is greater than in NO^+

Solution:

When bond order increases, bond length decreases. So

the bond length in NO is greater than in NO$^+$.

Hence option (4) is the answer.

CHEMICAL BONDING AND MOLECULAR STRUCTURE

Question 1. Isostructural species are those that have the same shape and hybridisation.

Among the given species, clarify the isostructural pairs.

(i) [NF_3 and BF_3]

(ii) [BF_4^- and NH_4^+]

(iii) [BCl_3 and $BrCl_3$]

(iv) [NH_3 and NO_3^-]

Answer 1. Option (ii) is the answer.

Question 2. Polarity in a molecule and hence the dipole moment depends primarily on

the constituent atoms' electronegativity and the molecule's shape. Which of

the following has the highest dipole moment?

(i) CO_2

(ii) HI

(iii) H_2O

(iv) SO2

Answer 2. Option (iii) is the answer.

Question 3. The types of hybrid orbitals of nitrogen in NO_2^+, NO_3^- and NH_4^+ respectively expected to be

(i) sp, sp3 and sp2

(ii) sp, sp2 and sp3

(iii) sp2, sp and sp3

(iv) sp2, sp3 and sp

Answer 3. Option (ii) is the answer.

Question 4. Hydrogen bonds are formed in many compounds, e.g., H2O, HF, and NH3. The boiling point of such compounds depends largely on the strength of the hydrogen bond and the number of the hydrogen bonds. The right decreasing order of the boiling points of the above compounds is :

(i) HF > H2O > NH3

(ii) H2O > HF > NH3

(iii) NH3 > HF > H2O

(iv) NH3 > H2O > HF

Answer 4. Option (ii) is the answer.

Question 5. In the PO_4^{3-} ion, formal charge on the oxygen atom of the P–O bond is

(i) + 1

(ii) – 1

(iii) – 0.75

(iv) + 0.75

Answer 5. Option (ii) is the answer.

Question 6. In NO3–ion, the number of the bond pairs and lone pairs of electrons on the nitrogen atom is

(i) 2, 2

(ii) 3, 1

(iii) 1, 3

(iv) 4, 0

Answer 6. Option (iv) is the answer.

Question 7. Which of the following species will have tetrahedral geometry?

(i) BH_4^-

(ii) NH_2^-

(iii) CO_3^{2-}

(iv) H_3O^+

Answer 7. Option (i) is the answer.

Question 8. The number of π bonds and σ bonds in the following structure is–

(i) 6, 19

(ii) 4, 20

(iii) 5, 19

(iv) 5, 20

Answer 8. Option (iii) is the answer.

Question 9. Which molecule/ ion out of the following does not contain unpaired electrons?

(i) N2+

(ii) O2

(iii) O22–

(iv) B2

Answer 09. Option (iii) is the answer.

Question 10. In which of the following molecules and ions all the bonds are not equal?

(i) XeF4

(ii) BF4–

(iii) C2H4

(iv) SiF4

Answer 10. Option (iii) is the answer.

Question 11. In which of the following substances will the hydrogen bond be strongest?

(i) HCl

(ii) H2O

(iii) HI

(iv) H2S

Answer 11. Option (ii) is the answer.

Question 12. If the electronic configuration of the element is 1s2 2s2 2p6 3s2 3p6 3d2 4s2, the

four electrons involved in the chemical bond formation will be_____.

(i) 3p6

(ii) 3p6, 4s2

(iii) 3p6, 3d2

(iv) 3d2, 4s2

Answer 12. Option (iv) is the answer.

Question 13. Which of the following angles corresponds to sp2 hybridisation?

(i) 90°

(ii) 120°

(iii) 180°

(iv) 109°

Answer 13. Option (ii) is the answer

Question 14. Which of the given formulas may represent the stable form of A:

(i) A

(ii) A2

(iii) A3

(iv) A4

Answer 14. Option (i) is the answer.

Question 15. Which of the given formulas may represent the stable form of C:

(i) C

(ii) C2

(iii) C3

(iv) C4

Answer 15. Option (ii) is the answer.

Question 16. The molecular formula of the compound formed from the B and C will be

(i) BC

(ii) B2C

(iii) BC2

(iv) BC3

Answer 16. Option (iv) is the answer.

Question 17. The bond in between B and C will be

(i) Ionic

(ii) Covalent

(iii) Hydrogen

(iv) Coordinate

Answer 17. Option (ii) is the answer.

Question 18. Which of the following order of the energies of molecular orbitals of N2 is correct?

(i) $(\pi 2py) < (\sigma 2pz) < (\pi^* 2px) \approx (\pi^* 2py)$

(ii) $(\pi 2py) > (\sigma 2pz) > (\pi^* 2px) \approx (\pi^* 2py)$

(iii) $(\pi 2py) < (\sigma 2pz) > (\pi^* 2px) \approx (\pi^* 2py)$

(iv) $(\pi 2py) > (\sigma 2pz) < (\pi^* 2px) \approx (\pi^* 2py)$

Answer 18. Option (i) is the answer.

Question 19. Which of the following statement is not right from the viewpoint of the molecular

orbital theory?

(i) Be2 is not a stable molecule.

(ii) He2 is not stable but He2+

is expected to exist.

(iii) Bond strength of N2 is the maximum amongst the homonuclear diatomic

molecules belonging to the second period.

(iv) The order of energies of the molecular orbitals in the N2 molecule is

σ2s < σ*2s < σ2pz < (π2px = π2py) < (π*2px = π*2py) < σ*2pz

Answer 19. Option (iv) is the answer.

Question 20. Which of the following options represents the correct bond order :

(i) O2– > O2 > O2+

(ii) O2– < O2 < O2+

(iii) O2– > O2 < O2+

(iv) O2– < O2 > O2

Answer 20. Option (ii) is the answer.

Question 21. Explain the non-linear shape of H2S and the non-planar shape of PCl3 using valence shell electron pair repulsion theory.

Answer 21.

In H2S, the Sulphur atom is surrounded by four electron pairs (two bond pairs as well as two lone pairs).

These four electron pairs adopt tetrahedral geometry.

The repulsion in between lone pair electrons brings distortion in the shape of the H2S.

So, H2S is not linear in shape.

Question 22. Using the molecular orbital theory, compare the bond energy and magnetic character of O2+ and O2− species.

Answer 22.

The Molecular Orbital configuration of O2+ as well as O-2 has given below:

O2+ (15): σ1s2 σ*1s2 σ2s2 σ*2s2 σ2pz2 π2px2 = π 2py2π*2px1

O2− (17): σ1s2 σ*1s2 σ2s2 σ*2s2 σ2pz2 π2px2 = π 2py2π*2px2= π*2py1

Bond order for the O2+ = 10-5/2 = 2.5

Bond order for the O-2 = 10-7/2 = 1.5

According to the Molecular Orbital Theory, the greater the bond order greater the bond energy.

Thus, O2+ is more stable than O2−

Question 23. Explain the shape of BrF5.

Answer 23.

BrF5's central atom is bromine, which has the hybridisation sp3d2.

Br atom has seven valence electrons, out of which five uses to make the pair with the F atoms, as well as two uses to make lone pairs of the electrons.

The lone pair and the bond pair repel each other. So, the shape is square Pyramidal.

Question 24. Structures of molecules of the two compounds are given below :

(a) Which of the following two compounds will have intermolecular hydrogen bonding, and which compound expects to show the intramolecular hydrogen bonding?

(b) The compound's melting point depends on, among other things, the extent of hydrogen bonding. Based on this, explain which of the above two compounds will show the higher melting point.

(c) The solubility of compounds in the water depends on the power to form hydrogen bonds with water. Which of the above compounds will easily form a hydrogen bond with water and be more soluble?

Answer 24.

(a) Compound 1 will be having intramolecular hydrogen bonding in o-nitrophenol

Compound (II) will have the intermolecular hydrogen bonding in p-nitrophenol.

(b) The compound (II) has a higher melting point because of the intermolecular bonding, a large number of molecules that will get attached.

(c) The compound (II) would be more soluble in water because it will easily form hydrogen bonding with the water molecules.

Question 25. Why does the type of overlap given in the following figure does not result in formation of bond?

Answer 25.

In figure (i), the area of the contact of ++ overlap is equal to the area of the +- overlap. The so-net overlap is zero.

In figure (ii), there is no overlap of the orbitals due to different symmetry.

Question 26. Explain why PCl_5 is trigonal bipyramidal,

whereas IF5 is square Pyramidal.

Answer 26.

In PCl5, P having the five valence electrons in the orbitals makes five bonds with 5 Cl atoms. It would share one of its electrons from the 3s to the 3d orbital. Therefore, the hybridisation will be sp3d, and the geometry will be trigonal bipyramidal.

IF5, the Iodine atom, has seven valence electrons in the molecular orbitals. It will form 5 bonds with the 5 Cl atoms using 5 electrons from its molecular orbital, and two electrons will form one lone pair on the Iodine atom, which gives the square pyramidal geometry.

Question 27. In both water and the dimethyl ether (CH3 — O — CH3), the oxygen atom is the central atom and having the same hybridisation, yet they have different bond angles. Which one has a greater bond angle? Elaborate with a reason.

Answer 27.

Dimethyl ether will have a greater bond angle. There will be more repulsion in between bond pairs of CH3 groups attached in ether than between bond pairs of the hydrogen atoms attached to oxygen in the water.

Question 28. Write the Lewis structure of the following compounds and shows a formal charge

on each atom.

HNO3, NO2, H2SO4

Answer 28.

The formal charge has been calculated by
Formal charge = ½ [total no: of bonding as well as shared electrons]

The formal charge on oxygen with the single bond = 6-6-2/2 = -1

The formal charge on oxygen with the double bond 6-4-4/2 = 0

The formal charge on the nitrogen = 5-2-6/2 = 0

The formal charge on oxygen 1 and 4 = 6-4-4/2 = 0

The formal charge on oxygen 2 and 3 = 6-4-4/2 = 0

The formal charge on hydrogen 1 and 2 = 1-0-2/2 = 0

The formal charge on sulfur = 6-0-12/2 = 0

Question 29. The energy of the $\sigma 2p_z$ molecular orbital is greater than $\pi 2p_x$ as well as $\pi 2p_y$ molecular orbitals in the nitrogen molecule. Write the complete sequence of the energy levels in the increasing order of the energy in the molecule. Compares the relative stability and the magnetic behaviour of the following species:

$N_2, N_2^+, N_2^-, N_2^{2+}$

Answer 29.

General sequence of the energy level of the molecular orbital has

$\sigma 1s < \sigma^* 1s < \sigma 2s < \sigma^* 2s < \pi 2p_x = \pi 2p_y < \sigma 2p_z$

$N_2\ \sigma 1s^2\ \sigma^* 1s^2\ \sigma 2s^2\ \sigma^* 2s^2\ \pi 2p_x^2 = \pi 2p_y^2\ \sigma 2p_z^2$

$N_2^+\ \sigma 1s^2\ \sigma^* 1s^2\ \sigma 2s^2\ \sigma^* 2s^2\ \pi 2p_x^2 = \pi 2p_y^2\ \sigma 2p_z^1$

$N_2^-\ \sigma 1s^2\ \sigma^* 1s^2\ \sigma 2s^2\ \sigma^* 2s^2\ \pi 2p_x^2 = \pi 2p_y^2\ \sigma 2p_z^2\ \sigma 2p_x^2$

$N_2^{2+}\ \sigma 1s^2\ \sigma^* 1s^2\ \sigma 2s^2\ \sigma^* 2s^2\ \pi 2p_x^2 = \pi 2p_y^2$

thus, Bond order = [(½)electrons in BMO − (½)electrons in the ABMO]

For N_2 = 10-4/2 = 3

Bond order for $N_2^+ = 9-4/2 = 2.5$

Bond order for $N_2^- = 10-5/2 = 2.5$

Bond order for $N_2^{2+} = 8-4/2 = 2$

so, the order of stability is:

$N_2 > N_2^- > N_2^+ > N_2^{2+}$

Question 30. What is the effect of the following process on the bond order in N_2 and O_2?

(i) $N_2 \rightarrow N_2^+ + e^-$

(ii) $O_2 \rightarrow O_2^+ + e^-$

Answer 30.

(i) N_2 will have 14 electrons when it donates one electron. These electrons remove from Bonding molecular orbital. as BO for $N_2 = 3$

(ii) O_2 has 16 electrons, 8 electrons in the molecular orbitals and 4 in the antibonding molecular orbitals.

BO for $O_2 = 2$

Question 31. Give reasons for the following :

(i) Covalent bonds are directional bonds, while ionic bonds are nondirectional.

(ii) The water molecule has a bent structure, whereas the carbon dioxide molecule is linear.

(iii) Ethyne molecule is linear.

Answer 31.

(i) A covalent bond is formed by overlapping atomic orbitals. The direction of the overlapping gives the direction of the bond.

(ii) In the water molecule, the oxygen atom is sp3 hybridised and has two lone pairs of electrons.

(iii) In the ethyne molecule, both of the carbon atoms are sp hybridised. The two sp hybrid orbitals of both of the carbon atoms orient in the opposite direction as, forming an angle of 180°.

Question 32. What is the ionic bond? With two suitable examples, explain the difference in between an ionic and the covalent bond?

Answer 32.

When the positively charged ion forms a bond with a negatively charged ion, one atom transfers electrons to another. An example of the ionic bond is the chemical compound Sodium Chloride (NaCl).

The difference between an ionic bond and a covalent bond is that an ionic bond essentially donates the electron to the other atom participating in the bond. In contrast, electrons in a covalent bond are shared equally between the atoms.

Question 33. Arrange the following of the bonds in order to increase the ionic character, giving a reason

N—H, F—H, C—H and O—H

Answer 33.

The ionic character is greater in the molecules with the highest electronegativity difference because the electron pair shifts toward the more electronegative atom, increasing the ionic character.

Thus, the ionic character order will be:

C-H < N-H < O-H < F-H

Question 34. Explain why CO_3^{2-} the single Lewis structure cannot represent an ion. How can it be best represented?

Answer 34.

The carbonate ion is represented in the form of a resonating hybrid structure. These structures are equivalent. In resonance, all three C-O bonds get a double character in one of the resonating structures.

So, all the bonds are equivalent and have equal lengths; hence carbonate ions cannot be represented by a single Lewis structure.

Question 35. Predict the hybridisation of each carbon in the molecule of the organic compound given below. Also, indicate this molecule's total number of sigma and pi bonds.

Answer 35.

The hybridisation of Carbon 1 is sp, carbon 2 is sp, carbon 3 sp2, carbon 4 is sp3, and carbon 5 has sp2.

The triple bond has two pi bonds as well as one sigma bond.

Each double bond has one sigma as well as one pie bond.

Every single bond is a sigma bond.

So, the total number of sigma bonds is 11 and pi bonds are 4 in the molecule.

Question 36. Group the following as linear as well as non-linear molecules :

H2O, HOCl, BeCl2, Cl2O

Answer 36.

BeCl2 has a linear structure

HOC1 is also non-linear in structure.

H2O has a V-shaped structure.

Cl_2O has a V-shaped structure.

Question 37. Elements X, Y and Z have 4, 5 and 7 valence electrons, respectively.

(i) Write the molecular formula for the compounds formed by these elements individually with the Hydrogen.

(ii) Which of these compounds would have the highest dipole moment?

Answer 37.

(i); XH_4, H_3Y, and HZ

Hydrogen has only one electron in its outermost shell. It shares one electron to form the covalent bond or accepts or donates one electron to form an ionic bond.

(ii) The compound HZ has a linear shape, and the difference in the electronegativity of Hydrogen and element Z is maximum.

Question 38. Draw the resonating structure of

(i) Ozone molecule

(ii) Nitrate ion

Answer 38.

Question 39. Predict the shape of the following molecules based on the hybridisation.

BCl_3, CH_4, CO_2, NH_3
Answer 39.

In compound BCl_3, Boron has sp2-hybridisation, and the shape is the Triangular Planar.

In methane CH_4, Carbon has sp3 -hybridisation and its shape are Tetrahedral.

In carbon dioxide CO2, carbon has sp-hybridisation, and its shape is Linear.

In ammonia NH3, nitrogen has sp3-hybridisation, and its shape is Pyramidal.

Question 40. All the C-O bonds in the carbonate ion (CO_3^{2-}) are the same length. Explain

Answer 40.

The carbonate ion is represented in the form of a resonating structure. These structures are equivalent to nature. In resonance, all 3 C-O bonds get a double character in one of the resonating structures.

Question 41. What is meant by the term average bond enthalpy? Why is the O—H bond enthalpy difference between ethanol (C2H5OH) and water?

Answer 41.

Similar bonds in the molecule do not have the same bond enthalpies. Mainly in the term average bond enthalpy is used in the polyatomic molecules. It is obtained by dividing the bond dissociation enthalpy by the number of bonds broken. The bond enthalpy of the OH bond is different in ethanol and water because of the difference in electronegativity of Hydrogen and carbon. As electronegativity differs in the Hydrogen and oxygen is higher than that in carbon and oxygen, so the O-H bond in the water has more bond enthalpy than in ethanol.

Question 42. Explain the formation of a chemical bond

Answer 42.

"Chemical bonds are an attractive force that binds the constituents of the chemical species together."

Thus, many theories have been suggested for chemical bond formation, such as the valence shell electron pair repulsion theory, electronic theory, molecular orbital theory as well as valence bond theory.

The formation of the chemical bond credits the system's tendency to achieve stability. It noticed that the inertness of the noble gases directly resulted from their filled outermost orbitals. Consequently, that proposed that the elements having a deficiency of the electrons in outermost shells are unstable. Thus, atoms combine and finish their separate octets or duplets to achieve the stable configuration of the closest inert gases. Thus, this combination may occur either by sharing the electrons. Thus, The formed chemical bond results from sharing the electrons among atoms and is known as a covalent bond. Also, a formed ionic bond results from sharing of electrons among atoms.

Question 43.

In both water and dimethyl ether (CH_3–O–CH_3), the oxygen atom is the central atom and has the same hybridisation, yet they have different bond angles. Which one has a greater bond angle? Specify with a reason.

Answer 43

The bond angle of dimethyl ether would be greater. More repulsion will exist in between bond pairs of the CH_3 groups attached in ether than in between bond pairs of hydrogen atoms attached to the oxygen in the water.

The carbon of the CH_3 in ether is attached to three hydrogen atoms via bonds, and these bonds' electron pairs contribute to the carbon atom's electronic charge density. As a result, the repulsion in between two CH_3 groups will be greater than between two hydrogen atoms.

Question 44. Explain why the BeH2 molecule has a zero dipole moment although the Be–H bonds are polar.

Answer 44.

Lewis structure of BeH2 is:

The central atom has no lone pair but two bond pairs. Thus, its shape is AB2. i.e. Linear shape.

so, the dipole moment of Be- H bond is equal and opposite in direction, nullifying one another. So, the dipole moment of BeH2 is 0.

Question 45. Which out of NH3 and NF3 has the higher dipole moment and why?

Answer 45.

N- atom is the central atom of the NF3 and NH3.

The central atom has one lone pair and three bond pairs. So, for both, the shape is AB3E. i.e. Pyramidal.

As, F-atom has more electronegativity than the H– atom, NF3 should have a higher dipole moment than the NH3. But the dipole moment of the NH3 is 1.46D which is higher than the dipole moment of NF3, which is 0.24D.

It gets clear from the directions of the dipole moments of individual bonds in NF3 and NH3.

As both the N-H bond is in the same direction, it adds to the bond moment of the lone pair, while N-F bonds are in the opposite direction, so they partly cut the bond moment of the lone pair.

Thus, the dipole moment of NH3 is higher than that of NF3.

Question 46. What is meant by the hybridisation of atomic orbitals? Describe the shapes of sp, sp2, and sp3 hybrid orbitals.

Answer 46.

"Hybridisation is the intermixing of a set of atomic orbitals of slightly different energies, forming a new set of the orbitals with equivalent energies and shapes".

E.g. 1 s- orbital hybridises with the 3 p- orbitals to form the 4 sp3 hybrid orbitals.

(a) sp hybrid orbital

1 s- orbital hybridises with 1 p- orbitals to form 2 sp hybrid orbitals. Sp hybrid orbital has a linear shape. The formation of sp orbital is:

(b) sp2 hybrid orbital

here the 1 s- orbital hybridises with 2 p- orbitals in order to form 3 sp2 hybrid orbitals. The shape of the sp2 orbital is trigonal planar.

(c) sp3 hybrid orbital

here the 1 s- orbital hybridises with 3 p- orbitals in order to form 4 sp3 hybrid orbitals. The shape of the sp3 orbital is the tetrahedron.

Question 47. Is there any change in the B and N atoms hybridisation due to the following reaction?

BF3 + NH3 —> F3B.NH3

Answer 47.

N- atom in NH3 has sp3 hybridisation. The orbital picture of N- the atom shows below:

B- atom in NF3 has sp2 hybridisation. The orbital picture of B- the atom shows below:

On the reaction between NH3 and BF3, F3B.NH3 is obtained

as a product, as hybridisation of B-atom changed to sp3. Although, the hybridisation of N- atoms remains unchanged.

Question 48. Which hybrid orbitals are used by the carbon atoms in the following molecules?

(a) CH3-CH3; (b) CH3-CH=CH2; (c) CH3CH2-OH; (d) CH3-CHO; (e) CH3COOH.

Answer 48.

(a) CH3-CH3

Here, C1 and C2 have sp3 hybridisation.

(b) CH3-CH=CH2

Here, C3 and C2 have sp2 hybridisation and C1 has sp3 hybridisation.

(c) CH3-CH2-OH

Here, C1 and C2 have sp3 hybridisation.

(d) CH3-CHO

Here, C1 has sp3 hybridisation, and C2 has sp2 hybridisation.

(e) CH3COOH

Here, C1 has sp3 hybridisation, and C2 has sp2 hybridisation.

Question 49. What do you understand by the bond pairs and lone pairs of electrons? Illustrate by giving one example of each type.

Answer 49.

A covalent bond forms when two atoms combine by sharing their valence electrons.

The shared pairs of electrons between the bonded atoms are called bond pairs. Every electron cannot participate in bonding. So the pairs of electrons which do not participate in bonding are called lone pairs.

for example, ethane has seven bond pairs but zero lone pairs.

b) Water has two bond pairs and two lone pairs on the O- atom.

Question 50. Distinguish between sigma and a pi bond.

Answer 50.

Sr. No.	Pi bond	Sigma bond
1	The lateral overlapping of orbitals forms a pi bond.	The end-to-end overlapping of orbitals forms the Sigma bond.
2	It is a comparatively weak bond.	It is a comparatively strong bond.
3	There is only one overlapping orbital, p-p.	The overlapping orbitals are s-s, s-p, and p-p.
4	Rotation around pi- the bond is restricted.	Rotation is possible around the sigma bond.
5	The electron cloud is not symmetrical about the line joining two nuclei.	The electron cloud is symmetrical about the line joining two nuclei.
6	It has 2 electron clouds, one above the plane of atomic nuclei and one below the plane of atomic nuclei.	It has 1 electron cloud, and that is symmetrical about the inter-nuclear axis.

Question 51. Explain the formation of the H2 molecule based on the valence bond theory.

Answer 51.

Let us assume 2 H- atoms X and Y with the nuclei NX and NY and electrons eX and eY, respectively.

When X and Y are far from each other, they have no interaction. Thus, as they come closer, the attractive force, as well as a repulsive force, become active.

The repulsive forces will be acting:

(a) Between electrons of both the atoms, eX and eY.

(b) Between nuclei of both the atoms, NX and NY.

The attractive forces will be acting:

(a) Between the electron and nucleus of the same atom, NX – eX and NY -eY.

(b) Between the electron of one atom and the nucleus of other atoms, NX–eY and NY–eX.

The repulsive force pushes the two atoms apart, whereas the attractive force tends to bring them together.

Repulsive forces:

Attractive forces:

The values of repulsive forces are lesser than the attractive forces. Thus, two atoms will approach each other. Therefore, there is a decrease in potential energy. In the end, a stage will be reached when the repulsive forces will balance the attractive forces, and the overall system achieves minimum energy, which leads to the formation of the H2 molecule.

Question 52. Write the important conditions that are required for the linear combination of the atomic orbitals to form the molecular orbitals.

Answer 52.

The condition that is required for the linear combination of atomic orbitals to form the molecular orbitals are as follows:

(i) The joining of atomic orbitals might have approximately the same energy. This implies in the homo-nuclear molecule, the 1s-orbital of one atom can join with the 1s- orbital of another atom but can't join with the 2s-orbital.

(ii) The joining atomic orbitals might have legitimate

orientations to ensure the maximum overlap.

(iii) The overlapping might be to a large extent.

Question 53. Write the significance of a plus as well as a minus sign shown in representing the orbitals

Answer 53.

Generally, the molecular orbital is represented by the 'wave function'.

A positive (+) sign representing a molecular orbital indicates a positive wave function.

A negative (-) sign representing a molecular orbital indicates a negative wave function.

Question 54. Define the octet rule. Write its significance and limitations

Answer 54.

The octet rule says, "atoms can combine either by transfer of valence electrons from one atom to another or by sharing the valence electrons to achieve the nearest inert gas configuration by having an octet in their valence shell."

The octet rule explains chemical bond formation depending upon the nature of the element.

Limitations:

(a) Octet rules fail to predict the relative stability as well as the shape of the molecules.

(b) It is based on the inert nature of the noble gases. But, some inert gases, say, krypton(Kr) as well as xenon(Xe), form compounds like KrF_2, XeF_2 etc.

(c) The octet rule can't apply to elements beyond the 3rd period. Elements present beyond the 3rd period have more than eight valence electrons surrounding the central atom. E.g. SF6, PF6 etc.

(d) The octet rule is not applied to atoms in a molecule with an odd number of electrons. E.g. For NO2 as well as NO octet rule is not applicable.

(e) If a compound has less than 8 electrons surrounding the central atom, then the octet rule can't be applied to that compound. E.g. BeH2, AlCl3, LiCl etc., have not to obey the octet rule.

Question 55. Write favourable factors for the formation of the ionic bond.

Answer 55.

An ionic bond forms by transferring one or more electrons from one atom to another. So, ionic bond formation depends on the flexibility of the neutral atoms to lose or gain electrons. The formation of an ionic bond also depends on the compound's lattice energy.

The factors that are favourable for the ionic bond formation:

(a) High electron affinity of atoms of non-metal.

(b) The high lattice energy of the compound which is formed.

(c) Low ionisation enthalpy of an atom of metal.

lattice energy.

The factors that are favourable for the ionic bond formation:

(a) High electron affinity of atoms of non-metal.

(b) The high lattice energy of the compound which is formed.

(c) Low ionisation enthalpy of an atom of metal.

Question 56. Discuss the shape of the following molecule using the VSEPR model:

BeCl2, BCl3, SiCl4, AsF5, H2S, PH3

Answer 56.

BeCl2

The central atom has no lone pair but two bond pairs. Thus, its shape is AB2. i.e. Linear shape.

BCl3

The central atom has no lone pair but three bond pairs. Thus, its shape is AB3. i.e. Trigonal planar.

SiCl4

The central atom has no lone pair but 4 bond pairs. Thus, its shape is AB4. i.e. Tetrahedral.

AsF5

The central atom has no lone pair but 5 bond pairs. Thus, its shape is AB5. i.e. Trigonal bipyramidal.

H2S

The central atom has one lone pair and two bond pairs. Thus, its shape is AB2E. i.e. Bent shape.

PH3

The central atom has one lone pair and three bond pairs. Thus, its shape is AB3E. i.e. Trigonal bipyramidal.

Question 57. Although geometries of NH3 and H2O molecules have distorted tetrahedral, the bond angle in water is less than that of Ammonia. Discuss.

Answer 57.

The geometry of H2O and NH3:

The central atom(N) in ammonia has one lone pair and three

bond pairs.

The water's central atom(O) has two lone pairs and two bond pairs.

So, these two lone pairs on O- atom in the water molecule repels the two bond pairs. And this repulsion is between the lone pair and bond pair on O- the atom of H2O is stronger than the repulsion between the lone pair and bond pair on the N- atom of NH3.

So, the bond angle in H2O is less than NH3, even though they have a distorted tetrahedral structure.

Question 58. How do you express the bond strength in terms of the bond order?

Answer 58.

The bond strength presents the extent of bonding between two atoms while forming a molecule. As the bond strength increases, thus, the bond becomes stronger, and the bond order increases.

Question 59. Define Bond length.

Answer 59.

Bond length is defined as an equilibrium distance in between the nuclei of 2 bonded atoms in a molecule.

Question 60. H3PO3 could be represented by structures 1 and 2 shown below. Can these two structures be a canonical form of the resonance hybrid representing H3PO3? If not, give reasons for the same.

Answer 60.

In given structures, the position of atoms is changed, so we can't take the two given structures as a canonical form of resonance hybrid representing H3PO3.

Question 61. Use Lewis symbols to show the electron transfer between the following atoms to form cations and anions :

(i) K and S

(ii) Ca and O

(iii) Al and N.

Answer 61.

(i) K and S

Electronic configurations of S and K are:

S: 2, 8, 6

K: 2, 8, 8, 1

Here, it is clear that K has one more electron than the nearest inert gas. i.e. Ne, whereas S need 2 electrons to complete its octet. So, the transfer of electrons takes place in the following way,

(ii) Ca and O

Electronic configurations of O and Ca are:

O: 2, 6

Ca: 2, 8, 8, 2

Here, it is clear that Ca has two more electrons than the nearest inert gas. i.e. Ar, whereas O needs 2 electrons to complete its octet. So, the transfer of electrons takes place in the following way,

(iii) Al and N

Electronic configurations of N and Al are:

N: 2, 5

Al: 2, 8, 3

Here, it is clear that Al has three more electrons than the nearest inert gas. i.e. Ne, whereas N needs 3 electrons to complete its octet. So, the transfer of electrons takes place in the following way,

Question 62. Write the significance and applications of the dipole moment.

Answer 62.

Key significant points for the dipole moment are as below:

The molecule's shape can be determined by it. Symmetrical molecules, such as linear, will have zero dipole moments, whereas non-symmetrical molecules will take on varied shapes, such as angular, bent, etc.
The polarity of molecules will be determined. The polarity will be lesser if the dipole moment is lesser and vice versa.
One can conclude that if a molecule has zero dipole moment, then it is non-polar, and if it shows some polar character, it is non-polar.

Question 63. Define electronegativity. How does it differ from the electron gain enthalpy?

Answer 63.

Electronegativity can be defined as the ability of an atom in a chemical compound to attract a bond pair of electrons towards itself.

Sr. No	Electronegativity	Electron affinity
1	Its electronegativity tends to attract the shared pairs of electrons for an atom in a chemical compound.	A tendency to attract and gain electrons for an isolated gaseous atom is its electron gain enthalpy.
2	It varies according to the element with which it is bonded.	It does not vary according to the element with which

it is bonded.

3   It is not constant for any element.   It is constant for an element.

4   It is not a measurable quantity.   It is a measurable quantity.

Question 64. Explain with the help of the suitable example polar covalent bond.

Answer 64.

When two unique atoms with distinct electronegativities join to form a covalent bond, the bond pair of electrons are not shared equally. The nucleus of the atom which has greater electronegativity will attract the bond pair. So, in this case the electron distribution gets distorted, and an electronegativity atom will attract the electron cloud.

Therefore, the electronegative element will get slightly negatively charged, and on the other side, the other atom will get slightly positively charged. As a result of this, there are two opposite poles developed in a molecule, and this type of bond formed is termed as a 'polar covalent bond'.

For e.g. HCl has a polar covalent bond. In HCl, Cl- atom has more electronegativity than the H- atom. So the bond pair shifts towards the Cl- atom, which acquires a positive charge.

Question 65. Arrange the bonds in order of increasing ionic character in the molecules: LiF, K2O, N2, SO2, and ClF3.

Answer 65.

The ionic character of a molecule depends on the difference in electronegativity between constituents atoms. So, the higher the difference, the higher the ionic character of a molecule.

So, the required sequence of ionic character of the above-given molecules is

N2< SO2< ClF3< K2O <LiF.

Question 66. As shown below, the skeletal structure of CH3COOH is correct, but some of the bonds are shown incorrectly. Write the correct Lewis structure for acetic acid.

Answer 66.

Correct Lewis structure of CH3COOH is given below:

Question 67. Explain the shape of BrF5.

Answer 67.

In BrF5, the central atom Br is surrounded by the five bonded pairs and one lone pair. This forms the shape of a square Pyramidal.

Question 68. Structures of molecules of the two compounds have given below:

(a) Which of the following two compounds will have intermolecular hydrogen bonding, and which compound is expected to show the intramolecular hydrogen bonding?

(b) The compound's melting point depends on, among other things, the extent of the hydrogen bonding. On the basis, explain which of the above two compounds would show the higher melting point.

(c) Solubility of the compounds in water depends on the power to form hydrogen bonds with water, which of the above compounds will easily form a hydrogen bond with water and be more soluble in it.

Answer 68.

(a) Hence the NO2 and OH groups in compound (I) are close together, and the intramolecular hydrogen bonding will form (II). Compound (II) will show the intermolecular hydrogen bonding.

(b) Hence it forms intramolecular hydrogen bonds, and compound (II) has a higher melting point. As a result, more and more molecules have linked together via hydrogen bond formation.

(c) Due to the intramolecular hydrogen bonding, compound (I) can't form hydrogen bonds with the water and is less soluble in it, whereas compound (II) can form hydrogen bonds with the water more easily and is thus more soluble in water.

Question 69. Why does the type of overlap given in the following figure not result in bond formation?

Answer 69.

In the first figure, the (++) overlap equals to the (+-) overlap; thus, these cancel out, and the net overlap is zero.
Hence the two orbitals are perpendicular to each other in the second figure. No overlap is possible.

Question 70. Explain why PCl_5 is trigonal bipyramidal, whereas IF_5 is square Pyramidal.

Answer 70.

P is surrounded by the five bond pairs and no lone pairs in PCl_5, whereas the iodine atom surrounds the five bond pairs and one lone pair in IF_5, so the shape of PCl_5 is trigonal bipyramidal, and IF_5 is square Pyramidal.

Question 71. In both water and dimethyl ether (CH_3–O–CH_3), the oxygen atom is the central atom and has the same hybridisation, yet they have different bond angles. Which one has a greater bond angle? Specify with a reason.

Answer 71.

The bond angle of dimethyl ether would be greater. More repulsion will exist in between bond pairs of the CH_3 groups attached in ether than in between bond pairs of hydrogen atoms attached to the oxygen in the water.

The carbon of the CH3 in ether is attached to three hydrogen atoms via bonds, and these bonds' electron pairs contribute to the carbon atom's electronic charge density. As a result, the repulsion in between two CH3 groups will be greater than between two hydrogen atoms.

COMPETIVE CORNER

NEET

1. C-O bond length is minimum in

(a) CO_2

(b) CO_3^{2-}

(c) $HCOO^-$

(d) CO

Answer: (d)

2. Molecules are held together in a crystal by

(a) hydrogen bond

(b) electrostatic attraction

(c) Van der Waal's attraction

(d) dipole-dipole attraction

Answer: (c)

3. Sp³d² hybridization is present in [Co(NH$_3$)$_6^{3+}$], find its geometry

(a) octahedral geometry

(b) square planar geometry

(c) tetragonal geometry

(d) tetrahedral geometry

Answer: (a)

4. Find the molecule with the maximum dipole moment

(a) CH_4

(b) NH_3

(c) CO_2

(d) NF_3

Answer: (b)

5. MX$_6$ is a molecule with octahedral geometry. How many X – M – X bonds are at 180°?

(a) four

(b) two

(c) three

(d) six

Answer: (c)

6. Find the pair with sp² hybridisation of the central molecule

(a) NH_3 and NO_2^-

(b) BF_3 and NH_2^-

(c) BF_3 and NO_2^-

(d) NH_2^- and H_2O

Answer: (c)

7. The formal charge and P-O bond order in PO_4^{3-} respectively are

(a) 0.6, -0.75

(b) -0.75, 1.25

(c) 1.0, -0.75

(d) 1.25, -3

Answer: (b)

8. Which of the molecules does not have a permanent dipole moment?

(a) SO_3

(b) SO_2

(c) H_2S

(d) CS_2

Answer: (d)

9. pπ – dπ bonding is present in which molecule

(a) SO_3^{2-}

(b) CO_3^{2-}

(c) NO_3^-

(d) BO_3^{3-}

Answer: (a)

10. Which one has a pyramidal shape?

(a) SO_3

(b) PCl_3

(c) CO_3^{2-}

(d) NO_3^-

Answer: (b)

JEE

1. The bond dissociation energy of B–F in BF_3 is 646 kJ mol^{-1} whereas that of C–F in CF_4 is 515 kJ mol^{-1}. The correct reason for higher B–F bond dissociation energy as compared to that of C–F is

(1) Significant pπ – pπ interaction between B and F in BF_3 whereas there is no possibility of such interaction between C and F in CF_4.

(2) Lower degree of pπ – pπ interaction between B and F in BF_3 than that

between C and F in CF_4

(3) Smaller size of B-atom as compared to that of C-atom

(4) Stronger bond between B and F in BF_3 as compared to that between C and F in CF_4.

Solution:

Because of pπ – pπ back bonding in BF_3 molecule, all B-F bonds have partial double bond character.

Hence option (1) is the answer.

2. Among the following species which two have a trigonal bipyramidal shape?

(1) NI_3 (2) I_3^- (3) SO_3^{2-} (4) NO_3^-

(1) II and III

(2) III and IV

(3) I and IV

(4) I and III

Solution:

Let us find the hybridization (H) and shape of given species.

(1) For NI_3, H = ½ (5+3) = 8/2 = 4 → sp^3 hybridized state. It is trigonal pyramidal in shape.

(2) For I_3^-, H = ½ (7+2+1) = 10/2 = 5 → sp^3d hybridized state. It is linear in shape.

(3) For SO_3^{2-}, H = ½ (6+2) = 8/2 = 4 → sp^3 hybridized state.

It is trigonal pyramidal in shape.

(4) For NO_3^-, H = ½ (5+1) = 6/2 = 3 → sp^2 hybridized state. It is trigonal planar in shape.

Hence option (4) is the answer.

3. Using MO theory, predict which of the following species has the shortest bond length?

(1) O_2^-

(2) O_2^{2-}

(3) O_2^{2+}

(4) O_2^+

Solution:

Chemical species	O_2^-	O_2^{2-}	O_2^{2+}	O_2^+
Bond order	1.5	1	3	2.5

Therefore bond length order $O_2^{2-} > O_2^- > O_2^+ > O_2^{2+}$

Hence option (3) is the answer.

4. Among the following, the species having the smallest bond is :

(1) NO

(2) NO$^+$

(3) O$_2$

(4) NO$^-$

Solution:

The bond order of given molecules are:

NO = 2.5, NO$^+$ = 3, O$_2$ = 2, NO$^-$ = 2

Larger the bond order, the smaller the bond length.

NO$^+$ has the largest bond order 3.

Therefore, it will have the smallest bond.

Hence option (2) is the answer.

5. The hybridisation of orbitals of N atom in NO$_3^-$, NO$_2^+$, NH$_4^+$ are respectively:

(1) sp^2, sp^3, sp

(2) sp, sp^3, sp^2

(3) sp, sp^2, sp^3

(4) sp^2, sp, sp^3

Solution:

In NO_3, the central N atom has 3 bonding domains and zero lone pairs of electrons.

In NO_2, the central N atom has 2 bonding domains and zero lone pairs of electrons.

In NH_4, the central N atom has 4 bonding domains and zero lone pairs of electrons.

The Hybridization of N atom in NO_3^-, NO_2^+, NH_4^+ are sp^2, sp, sp^3 respectively.

Hence option (4) is the answer.

6. Based on lattice energy and other considerations, which one of the following alkali metal chlorides is expected to have the highest melting point?

(1) RbCl

(2) LiCl

(3) KCl

(4) NaCl

Solution:

NaCl has the highest melting point.

Hence option (4) is the answer.

7. The structure of IF_7 is :

(1) octahedral

(2) pentagonal bipyramid

(3) square pyramid

(4) trigonal bipyramid

Solution:

For IF_7, hybridisation – sp^3d^3. The shape is pentagonal bipyramidal.

Hence option (2) is the answer.

8. Which of the following has the square planar structure :

(1) NH_4^+

(2) CCl_4

(3) XeF_4

(4) BF_4^-

Solution:

Hybridization of XeF_4 sp^3d^2

It has a square planar shape.

Hence option (3) is the answer.

9. Among the following the maximum covalent character is shown by the compound :

(1) $AlCl_3$

(2) $MgCl_2$

(3) $FeCl_2$

(4) $SnCl_2$

Solution:

Al^{+3} is having the highest polarizing power than other compounds having greater covalent character.

Hence option (1) is the answer.

10. The compound of Xenon with zero dipole moment is :

(1) XeO_3

(2) XeO_2

(3) XeF_4

(4) $XeOF_4$

Solution:

XeF_4 has dipole moment zero.

Hence option (3) is the answer.

11. Which of the following has a maximum number of lone pairs associated with Xe?

(1) XeO_3

(2) XeF_4

(3) XeF_6

(4) XeF_2

Solution:

XeO_3 has 1 lone pair of electrons. XeF_4 has 2 lone pairs of electrons. XeF_6 has 1 lone pair of electrons. XeF_2 has 3 lone pairs of electrons. XeF_2 has a maximum number of lone pairs of electrons.

Hence option (4) is the answer.

CHEMISTRY PART 1

12. Among the following the molecule with the lowest dipole moment is :

(1) $CHCl_3$

(2) CH_2Cl_2

(3) CCl_4

(4) CH_3Cl

Solution:

The order of the dipole moment is $CCl_4 < CHCl_3 < CH_2Cl_2 < CH_3Cl$. So CCl_4 has the lowest dipole moment.

Hence option (3) is the answer.

13. The number of types of bonds between two carbon atoms in calcium carbide is

(1) One sigma, two pi

(2) One sigma, one pi

(3) Two sigma, one pi

(4) Two sigma, two pi

Solution:

$CaC_2 \rightarrow Ca^{+2} + C_2^{2-}$

$^-C \equiv C^-$

Number of sigma bond is 1 and number of pi bond is 2.

Hence option (1) is the answer.

14. The formation of molecular complex BF_3 – NH_3 results in a change in the hybridisation of boron

(1) From sp^3 to sp^3d

(2) From sp^2 to dsp^2

(3) From sp^3 to sp^2

(4) From sp^2 to sp^3

Solution:

In BF_3, Boron atom has 3 bond pairs of electrons and 0 lone pairs of electrons. It is sp^2 hybridized. In $F_3B \leftarrow NH_3$, Boron atom has 4 bond pairs of electrons and 0 lone pairs of electrons. It is sp^3 hybridized. So the formation of molecular complex results in a change in the hybridization of boron from sp^2 to sp^3.

Hence option (4) is the answer.

15. The molecule having the smallest bond angle is :

(1) PCl_3

(2) NCl_3

(3) $AsCl_3$

(4) $SbCl_3$

Solution:

Bond angle order $NCl_3 > PCl_3 > AsCl_3 > SbCl_3$.

Hence option (4) is the answer.

16. In which of the following pairs the two species are not isostructural?

(1) AlF_6^{3-} and SF_6

(2) CO_3^{2-} and NO_3^-

(3) PCl_4^+ and $SiCl_4$

(4) PF_5 and BrF_5

Solution:

PF_5 has a trigonal bipyramidal shape. BrF_5 has a square pyramidal shape.

Hence option (4) is the answer.

17. Which one of the following molecules is expected to exhibit diamagnetic behaviour?

(1) C_2

(2) N_2

(3) O_2

(4) S_2

Solution:

C_2 and N_2 have no unpaired electrons. So they exhibit diamagnetic behaviour.

18. Which of the following is the wrong statement?

(1) ONCl and ONO⁻ are not isoelectronic

(2) O_3 molecule is bent

(3) Ozone is violet-black in solid-state

(4) Ozone is diamagnetic gas

Solution:

In the given options all are correct statements.

19. Stability of the species Li_2, Li_2^- and Li_2^+ increases in the order of :

(1) $Li_2 < Li_2^+ < Li_2^-$

(2) $Li_2^- < Li_2^+ < Li_2$

(3) $Li_2 < Li_2 < Li_2^+$

(4) $Li_2^- < Li_2 < Li_2^+$

Solution:

The bond order of Li_2 is 1. The bond order of Li_2^+ is 0.5. The bond order of Li_2^- is 0.5. Stability will depend on the bond order. Li_2^+ is more stable than Li_2^- because the higher interelectronic repulsion in Li_2^- makes it the least stable. So the order is $Li_2 > Li_2^+ > Li_2^-$.

Hence option (2) is the answer.

20. In which of the following pairs of molecules/ions, both species are not likely to exist?

(1) H_2^+, He_2^{2-}

(2) H_2^-, He_2^{2-}

(3) H_2^{2+}, He_2

(4) H_2^-, He_2^{2+}

Solution:

The bond order of H_2^{2+} and He_2 is zero. So these molecules do not exist.

Hence option (3) is the answer.

21. Bond distance in HF is 9.17×10^{-11} m. Dipole moment of HF is 6.104×10^{-30} Cm. The per cent ionic character in HF will be : (electron charge = 1.60×10^{-19} C)

(1) 61.0%

(2) 38.0%

(3) 35.5%

(4) 41.5%

Solution:

Given Bond distance = 9.17×10^{-11} m.

Dipole moment = 6.104×10^{-30} Cm

% iconic character = $6.104 \times 10^{-30} \times 100 / (1.60 \times 10^{-19} \times 9.17 \times 10^{-11})$

= 41.5%

Hence option (4) is the answer.

22. In which of the following ionization processes the bond energy has increased and also the magnetic behaviour has changed from paramagnetic to diamagnetic?

(1) $NO \rightarrow NO^+$

(2) $O_2 \rightarrow O_2^+$

(3) $N_2 \rightarrow N_2^+$

(4) $C_2 \rightarrow C_2^+$

Solution:

During the ionisation of $NO \rightarrow NO^+$, the bond order changes from 2.5 to 3. Also magnetic character changes from paramagnetic to diamagnetic.

During the ionisation of $O_2 \rightarrow O_2^+$, the bond order increases from 2 to 2.5 and the magnetic character changes from paramagnetic to diamagnetic.

During the ionisation of $N_2 \rightarrow N_2^+$, the bond order decreases from 3 to 2.5 and the magnetic behaviour changes from diamagnetic to paramagnetic.

During the ionisation of $C_2 \to C_2^+$, the bond order decreases from 2 to 1.5 and the magnetic behaviour changes from diamagnetic to paramagnetic.

Hence option (1) is the answer.

23. Which one of the following molecules is paramagnetic?

(1) NO

(2) O_3

(3) N_2

(4) CO

Solution:

NO has an unpaired electron. So it is paramagnetic in nature.

Hence option (1) is the answer.

24. The catenation tendency of C, Si and Ge is in the order Ge < Si < C. The bond energies (in kJ mol^{-1} of C—C, Si—Si and Ge—Ge bonds are respectively :

(1) 348, 260, 297

(2) 348, 297, 260

(3) 297, 348, 260

(4) 260, 297, 348

Solution:

Bond energy order is C – C > Si – Si > Ge – Ge.

Hence option (2) is the answer.

25. Oxidation state of sulphur in anions SO_3^{2-}, $S_2O_4^{2-}$ and $S_2O_6^{2-}$ increases in the orders

(1) $S_2O_6^{2-} < S_2O_4^{2-} < SO_3^{2-}$

(2) $SO_3^{2-} < S_2O_4^{2-} < S_2O_6^{2-}$

(3) $S_2O_4^{2-} < SO_3^{2-} < S_2O_6^{2-}$

(4) $S_2O_4^{2-} < S_2O_6^{2-} < SO_3^{2-}$

Solution:

The oxidation state of sulphur in SO_3^{2-} is +4. The Oxidation state of sulphur in $S_2O_4^{2-}$ is +3 and in $S_2O_6^{2-}$ is +5. So the order is $S_2O_4^{2-} < SO_3^{2-} < S_2O_6^{2-}$

Hence option (3) is the answer.

26. In which of the following species is the underlined carbon having sp³ hybridisation?

(1) CH$_3$COOH

(2) CH$_3$CH$_2$OH

(3) CH$_3$COCH$_3$

(4) CH$_2$=CH–CH$_3$

Solution:

Only in CH$_3$CH$_2$OH, carbon has sp³ hybridisation.

In other molecules, the carbon atom has multiple bonds,

Hence option (2) is the answer.

27. In which of the following sets, all the given species are isostructural?

(1) BF$_3$, NF$_3$, PF$_3$, AlF$_3$

(2) PCl$_3$, AlCl$_3$, BCl$_3$, SbCl$_3$

(3) BF$_4^-$, CCl$_4$, NH$_4^+$, PCl$_4^+$

(4) CO$_2$, NO$_2$, ClO$_2$, SiO$_2$

Solution:

BF_4^-, CCl_4, NH_4^+, PCl_4^+ are tetrahedral.

Hence option (3) is the answer.

28. In XeF_2, XeF_4, XeF_6 the number of lone pairs of Xe are respectively

(1) 2, 3, 1

(2) 1, 2, 3

(3) 4, 1, 2

(4) 3, 2, 1

Solution:

XeF_2 has 3 lone pairs of electrons. XeF_4 has 2 lone pairs of electrons. XeF_6 has 1 lone pair of electrons.

Hence option (4) is the answer.

29. Which of the following statements is true?

(1) HF is less polar than HBr

(2) absolutely pure water does not contain any ions

(3) chemical bond formation take place when forces of

attraction overcome the forces of repulsion

(4) in covalency transference of electron takes place

Solution:

Chemical bond formation takes place when forces of attraction overcome the forces of repulsion.

Hence option (3) is the answer.

30. Which one of the following pairs of molecules will have permanent dipole moments for both members?

(1) NO_2 and CO_2

(2) NO_2 and O_3

(3) SiF_4 and CO_2

(4) SiF_4 and NO_2

Solution:

NO_2 and O_3 have angular shapes. So they will have a net dipole moment.

Hence option (2) is the answer.

31. The states of hybridization of boron and oxygen atoms in boric acid (H_3BO_3) are respectively

(1) sp^2 and sp^2 (2) sp^3 and sp^3

(3) sp^3 and sp^2 (4) sp^2 and sp^3

Solution:

Hybridization of B is sp^2 and O is sp^3

Hence option (4) is the answer.

32. The maximum number of 90° angles between bond pair of electrons is observed in

(1) dsp^3 hybridization

(2) sp^3d^2 hybridization

(3) dsp^2 hybridization

(4) sp^3d hybridization

Solution:

sp^3d^2 hybridisation has an octahedral configuration. All the bond angles are 90° in the structure.

Hence option (2) is the answer.

33. Which of the following are arranged in an increasing order of their bond strengths?

(1) $O_2^- < O_2 < O_2^+ < O_2^{2-}$

(2) $O_2^{2-} < O_2^- < O_2 < O_2^+$

(3) $O_2^- < O_2^{2-} < O_2 < O_2^+$

(4) $O_2^+ < O_2 < O_2^- < O_2^{2-}$

Solution:

Higher the bond order, stronger the bonds. The increasing order is $O_2^{2-} < O_2^- < O_2 < O_2^+$.

Hence option (2) is the answer.

34. Bond order and magnetic nature of CN⁻ are respectively

(1) 3, diamagnetic

(2) 2.5, paramagnetic

(3) 3, paramagnetic

(4) 2.5, diamagnetic

Solution:

Bond order = ½ [$n_b - n_a$]

= ½ [10-4]

= ½ (6)

= 3

It does not have unpaired electrons. So, it is diamagnetic.

Hence option (1) is the answer.

35. The bond order in NO is 2.5 while that in NO^+ is 3. Which of the following statements is true for these two species?

(1) Bond length in NO^+ is greater than in NO

(2) Bond length is unpredictable

(3) Bond length in NO^+ in equal to that in NO

(4) Bond length in NO is greater than in NO^+

Solution:

When bond order increases, bond length decreases. So the bond length in NO is greater than in NO^+.

Hence option (4) is the answer.

STATES OF MATTER

COMPETITIVE CORNER:

NEET

1. A container with a pin-hole contains equal moles of $H_{2(g)}$ and $O_{2(g)}$. Find the fraction of oxygen gas escaped at the same time when one-fourth of hydrogen gas escapes

(a) 1/16

(b) 1/4

(c) 1/2

(d) 1/8

Answer: (a)

2. What are the conditions for gas like Carbon monoxide to obey the ideal gas laws?

(a) low temperature and low pressure

(b) low temperature and high pressure

(c) high temperature and low pressure

(d) high temperature and high pressure

Answer: (c)

3. If the temperature is doubled, the average velocity of a gaseous molecule increases by

(a) 4

(b) 1.4

(c) 2

(d) 2.8

Answer: (b)

4. Find the molecular mass of a gas that takes three times more time to effuse as compared to He with the same volume

(a) 9 u

(b) 64 u

(c) 27 u

(d) 36 u

Answer: (d)

5. At the same temperature, the average molar kinetic energy of N_2 and CO is

(a) $KE_1 > KE_2$

(b) $KE_1 < KE_2$

(c) $KE_1 = KE_2$

(d) insufficient information given

Answer: (c)

6. Find the temperature at which the rate of effusion of N_2 is 1.625 times to that of SO_2 at 500°C

(a) 620°C

(b) 173°C

(c) 110°C

(d) 373°C

Answer: (a)

7. Find the change in the root mean square speed of the

gas on raising the temperature from 27°C to 927°C

(a) becomes times

(b) gets doubled

(c) gets halved

(d) remains same

Answer: (b)

8. Find the relation between probable velocity, mean velocity and root mean square velocity

(a)

(b)

(c)

(d)

Answer: (c)

9. If 1.204×10^{21} molecules of H_2SO_4 are removed from 392 mg of H_2SO_4, find the moles of H_2SO_4 left.

(a) 4×10^{-3}

(b) 1.5×10^{-3}

(c) 1.2×10^{-3}

(d) 2×10^{-3}

Answer: (d)

10. Find the fraction of the total pressure exerted by hydrogen if it is mixed with ethane in an empty container at 25°C

(a) 15/16

(b) 1/16

(c) 1/2

(d) 1

Answer: (a)

JEE

1. If Z is the compressibility factor, van der Waals'

equation at low pressure can be written as :

(1) $Z = 1 - Pb/RT$

(2) $Z = 1 + Pb/RT$

(3) $Z = 1 + RT/Pb$

(4) $Z = 1 - a/V_m RT$

Solution:

$(P + a/V_m^2)(V_m - b) = RT$ [Van der waals equation of state]

$V_m - b \approx V_m$

So the equation becomes $(P + a/V_m^2)V_m = RT$

$\Rightarrow PV_m + a/V_m = RT$

Divide all terms by RT

$PV_m/RT + a/V_m RT = RT/RT$

$PV_m/RT = 1 - a/V_m RT$

$Z = 1 - a/V_m RT$ [$Z = PV_m/RT$]

Hence option (4) is the answer.

2. Which intermolecular force is most responsible in allowing xenon gas to liquefy?

(1) Instantaneous dipole induced dipole

(2) Ion dipole

(3) Ionic

(4) Dipole-dipole

Solution:

For the liquefaction of xenon, instantaneous dipole induced dipole forces are responsible.

Hence option (1) is the answer.

3. The temperature at which oxygen molecules have the same root mean square speed as helium atoms have at 300 K is :

(Atomic masses : He = 4 u, O = 16 u)

(1) 1200 K

(2) 600 K

(3) 300 K

(4) 2400 K

Solution:

Given Atomic masses : He = 4 u, O = 16 u

$(V_{rms})\,O_2 = (V_{rms})\,He$

$\sqrt{3RT_1/M_1} = \sqrt{3RT_2/M_2}$

$T_1/M_1 = T_2/M_2$

$T_1/32 = 300/4$

$T_1 = 300 \times 32/4$

= 2400 K

Hence option (4) is the answer.

4. The compressibility factor for a real gas at high pressure is :

(1) 1−Pb/RT

(2) 1+ RT/Pb

(3) 1

(4) 1+Pb/RT

Solution:

$(P+a/V^2)(V-b) = RT$ [Real gas equation]

a/V^2 can be neglected at high pressure.

$PV - Pb = RT$

$PV/RT = (RT/RT) + (Pb/RT)$

$PV/RT = 1 + (Pb/RT)$ ...(1)

$Z = PV/RT$ ...(2)

Equating (1) and (2)

$Z = 1 + (Pb/RT)$

Hence option (4) is the answer.

5. The relationship among most probable velocity, average velocity and root mean Square velocity is respectively :

(1) $\sqrt{2} : \sqrt{(8/\pi)} : \sqrt{3}$

(2) $\sqrt{2} : \sqrt{3} : \sqrt{(8/\pi)}$

(3) $\sqrt{3} : \sqrt{(8/\pi)} : \sqrt{2}$

(4) $\sqrt{(8/\pi)} : \sqrt{3} : \sqrt{2}$

Solution:

$V_{mpv} = \sqrt{(2RT/M)}$

$V_{av} = \sqrt{(8RT/\pi M)}$

$V_{rms} = \sqrt{(3RT/M)}$

$V_{mpv} : V_{av} : V_{rms} = \sqrt{(2RT/M)} : \sqrt{(8RT/\pi M)} : \sqrt{(3RT/M)}$

$= \sqrt{2} : \sqrt{(8/\pi)} : \sqrt{3}$

Hence option (1) is the answer.

6. Value of gas constant R is

(1) 0.082 L atm

(2) 0.987 cal mol^{-1} K^{-1}

(3) 8.3 J mol^{-1} K^{-1}

(4) 83 erg mol^{-1}K^{-1}

Solution:

R = 8.3 J mol^{-1} K^{-1}

Hence option (3) is the answer.

7. By how many folds the temperature of a gas would increase when the root mean

Square velocity of the gas molecules in a container of fixed volume is increased from 5×10^4 cm/s to 10×10^4 cm/s?

(1) Four

(2) three

(3) Two

(4) Six

Solution:

$V_{rms} \propto \sqrt{T}$

$V_1/V_2 = \sqrt{(T_1/T_2)} = 5 \times 10^4 / 10 \times 10^4$

squaring, we get

$T_1/T_2 = 25/100 = 1/4$

$T_2 = 4T_1$

Hence option (1) is the answer.

8. Kinetic theory of gases proves

(1) Only boyle's law

(2) Only Charle's law

(3) Only Avogadro's law

(4) all of these

Solution:

One of the postulates of kinetic theory of gases is average kinetic energy proportional to T.

This theory proves all the above given laws.

Hence option (4) is the answer.

9. Which one of the following is the wrong assumption of kinetic theory of gases?

(1) All the molecules move in a straight line between collision and with the same velocity.

(2) Molecules are separated by great distances compared to their sizes.

(3) Pressure is the result of elastic collision of molecules with the container's wall.

(4) Momentum and energy always remain conserved.

Solution:

The molecules are always in random motion and obey Newton's law of motion. They have velocities in all directions ranging from zero to infinity.

Hence option (1) is the answer.

10. 'a' and 'b' are Vander Waal's constants for gases. Chlorine is more easily liquified than ethane because:

(1) a for Cl_2 < a for C_2H_6 but b for Cl_2 > b for C_2H_6

(2) a for Cl_2 > a for C_2H_6 but b for Cl_2 < b for C_2H_6

(3) a and b for Cl_2 > a and b for C_2H_6

(4) a and b for Cl_2 < a and b for C_2H_6

Solution:

Greater the 'a' value, more easily the gas is liquified, lower the 'b' value, more easily the gas is liquified.

Hence option (2) is the answer.

11. A gaseous compound of nitrogen and hydrogen contains 12.5% (by mass) of Hydrogen. The density of the compound relative to hydrogen is 16. The molecular formula of the compound is :

(1) NH_2

(2) NH_3

(3) N_3H

(4) N_2H_4

Solution:

Given that gaseous compound of nitrogen and hydrogen contains 12.5% (by mass) of

Hydrogen.

Element	Percentage	Atomic ratio	Simple ratio
H	12.5%	12.5/1 = 12.5	12.5/6.25 = 2
N	87.5%	87.5/14 = 6.25	6.25 / 6.25 = 1

Empirical formula = NH_2

Empirical mass = 16

Molecular weight = 2× Vapour density = 2×16 = 32

So n = molecular mass / empirical mass = 32/16 = 2

Molecular formula = Empirical formula × n

= $(NH_2) \times 2$

= N_2H_4

Hence option (4) is the answer.

12. The one that is extensively used as a piezo electric material is

(1) quartz

(2) amorphous silica

(3) trydymite

(4) mica

Solution:

Quartz is used as a piezo electric material.

Hence option (1) is the answer.

13. Which primitive unit cell has unequal edge lengths and all axial lengths different from 90^0.

(1) Monoclinic

(2) Triclinic

(3) Tetragonal

(4) Hexagonal

Solution:

Triclinic primitive unit cell has unequal edge lengths and all axial lengths different from $90°$.

Hence option (2) is the answer.

14. In Van der waals equation of state of the gas law, the constant b is a measure of

(1) Intermolecular repulsions

(2) Intermolecular attraction

(3) Volume occupied by molecules

(4) Intermolecular collisions per unit volume.

Solution:

Van der waals constant b is the measure of effective volume occupied by the gas molecules.

Hence option (3) is the answer.

15. A pressure cooker reduces cooking time for food because

(1) Heat is more evenly distributed in the cooking space.

(2) B.P of water involved in cooking is increased

(3) The higher pressure inside the cooker crushes the

food.

(4) Cooking involves chemical changes helped by a rise in temperature.

Solution:

By Gay Lussac's law, at constant pressure of a given mass of a gas is directly proportional to the absolute temperature of the gas. So on increasing pressure, temperature also increases. So the boiling point of water is also increased.

Hence option (2) is the answer.

16. According to kinetic theory of gases, in an ideal gas, between two successive collisions a gas molecule travels

(1) In a circular path

(2) In wavy path

(3) In a straight line path

(4) with an accelerated velocity

Solution:

According to kinetic theory of gases, in an ideal gas, between two successive collisions a gas molecule travels in a straight line path.

Hence option (3) is the answer.

THERMODYNAMICS

Question 1: The thermodynamic state function is the quantity

(i) used to determine the heat changes

(ii) whose value is independent of the path

(iii) used to determine the pressure, volume and work

(iv) whose value is dependent on the temperature only

Answer 1: (ii) A quantity that is independent of the path.

Explanation: Functions like the pressure, volume and temperature are dependent on the state of the system only, not on the path.

Question 2: For the process to happen under the adiabatic conditions, the correct condition can be given by:

(i) $\Delta T = 0$ (ii) $\Delta p = 0$

(iii) $q = 0$ (iv) $w = 0$

Answer 2: (iii) $q = 0$

Explanation: For the adiabatic process, the heat transfer is zero, that is, $q = 0$.

Question 3: The enthalpies for all the elements in their standard states are given as:

(i) Unity

(ii) Zero

(iii) < 0

(iv) Different from each of the element

Answer 3: (ii) Zero

Question 4. 18.0 g for the water completely evaporates at 100°C as well as 1 bar pressure, as well as the enthalpy change for the process is 40.79 kJ mol-1. What would be the enthalpy change for vaporising the two moles of the water under the given conditions? What is the standard enthalpy for the vaporisation of the water?

Answer 4. Given condition,

The amount of the water is 18.0 g, as well as the pressure, is 1 bar.

Now, we know that 18.0 g H2O = 1 mole H2O.

Enthalpy change of vaporising the 1 mole of H2O = 40.79 kJ mol-1

So, Enthalpy change of vaporising the 2 moles of H2O = 2 × 40.79 kJ = 81.358kJ

Standard enthalpy for evaporation at 100°C and 1 bar pressure is given as,

Δvap H2O = + 40.79 kJ mol-1.

Question 5. One mole of the acetone needs less heat to evaporate as compared to the 1 mol of the water. Which of the following two liquids has higher enthalpy for the vapourisatIon?

Answer 5. Acetone needs less heat to vaporise because of the weak force of attraction between the molecules.

As a result, the water has a higher enthalpy of evaporation.

It can be given as:

Δvap(water) > Δvap(acetone)

Question 6. Standard molar enthalpy for the formation, $\Delta_f H^\ominus$, is just the special case of the enthalpy of the reaction, $\Delta_r H^\ominus$. Is the $\Delta_r H^\ominus$ for the following reaction equal to $\Delta_f H^\ominus$? Give proper reasons for your answer.

$CaO(s) + CO_2(g) \rightarrow CaCO_3(s); \Delta_f H^\ominus = -178.3$ kJ mol-1

Answer 6. No, the $\Delta_r H^\ominus$ for the given reactions is not equal to $\Delta_f H^\ominus$.

The standard enthalpy change to the formation of the one mole of the compound from the elements in their most stable states is known as the standard molar enthalpy for formation, $\Delta_r H^\ominus$.

$CaO(s) + C(s) + 3/2\ O_2(g) \rightarrow CaCO_3(s)$

This reaction is not the same as the given reaction.

Thus, $\Delta_r H^\ominus \neq \Delta_f H^\ominus$.

Question 7. The value for the $\Delta_f H^\ominus$ of the NH_3 is

-91.8 KJ mol-1. Calculate the enthalpy change for the given reaction:

$2NH_3(g) \rightarrow N_2(g) + 3H_2(g)$

Answer 7. $\Delta_f H^\ominus$ is given for the formation. For the given reverse reaction, $\Delta_f H^\ominus$ changes sign due to the opposite of the exothermic reaction in the endothermic reaction.

Thus, for the given one mole, $\Delta_f H^\ominus$ for the decomposition is – (-91.8) = 91.8.

But the two moles are decomposing here.

Hence,

$\Delta_f H^\ominus = 2 \times 91.8 = 183.6$ KJ mol-1.

Question 8. Enthalpy is an extensive property. In general, if the enthalpy for the overall reaction A→B along one route is considered as ΔrH and ΔrH1, ΔrH2, ΔrH3 ..... represent enthalpies of the intermediate reactions leading to the formation of the product B. What will be the relation between the ΔrH for the overall reaction as well as the ΔrH1, ΔrH2 ..... etc., for the intermediate reactions?

Answer 8. In general, if the enthalpy of the overall reaction A→B along one route is considered as ΔrH and ΔrH1, ΔrH2, ΔrH3 ..... represent enthalpies of reaction leading to the formation of the same product B through the other route,

So, we get, ΔrH = ΔrH1 + ΔrH2 + ΔrH3 +.....

For the general reaction, Hess's law of the constant heat addition can be given as.

Question 9. The enthalpy of the atomisation for the reaction $CH_4(g) \to C(g) + 4H(g)$ is 1665 kJ mol-1. What is the bond of energy of the C–H bond?

Answer 9. Enthalpy of the atomisation for the 4 moles of the C–H bonds = 1665 kJ mol-1

Hence, C–H bond energy for given per mole = 1665 kJ mol-1/4 = 416.2 kJmol-1.

Question 10. Use the given data to calculate the lattice Δ H⊖ for NaBr.

Δsub H⊖ for the sodium metal = 108.4 kJ mol-1

Ionisation enthalpy of the sodium = 496 kJ mol-1

Electron gain enthalpy for the bromine = – 325 kJ mol-1

Bond dissociation enthalpy for the bromine = 192 kJ mol-1

ΔfH⊖ of the NaBr (s) = – 360.1 kJ mol-1

Answer 10. According to Hess's Law,

$\Delta_f H^\ominus = \Delta_{sub} H^\ominus + \Delta_{IE} H^\ominus + \Delta_{diss} H^\ominus + \Delta_{tg} H^\ominus + U$

$\Delta_{sub} H^\ominus$ of the Na metal = 108.4 kJ/mol

Ionisation Enthalpy of Na = 496 k/mol

$\Delta_{tg} H^\ominus$ for Br = – 325 kJ mol–1

$\Delta_{diss} H^\ominus$ for Br = 192 kJ mol–1

$\Delta_f H^\ominus$ of NaBr = – 360.1 kJ mol–1

$\Delta_f H^\ominus = \Delta_{sub} H^\ominus$ +Ionisation Enthalpy of Na + $\Delta_{diss} H^\ominus$ + $\Delta_{tg} H^\ominus$ + U

-360.1 = 108.4 +496 + 96 + (-325) – U

U = +735.5 kJ/mol

Question 11. Given the condition, $\Delta H = 0$ for the mixing of the two gases. Explain if the diffusion of these gases into each other in the closed container is a spontaneous process or not.

Answer 11. The diffusion will be a spontaneous process. As the change in the enthalpy is zero, the change in the randomness or disorder that is ΔS increases. As a result, for the equation $\Delta G = \Delta H - T\Delta S$, the term $T\Delta S$ will be given as negative. Thus, ΔG will be negative. Hence the process will be spontaneous.

Question 12. Heat has a randomising influence on the system as well as temperature is the measure of the average chaotic motion of the particles in the system. Write the mathematical relation which relates to these three parameters.

Answer 12. Heat has a randomising influence on the system as well as temperature is the measure of the average chaotic motion of the particles in the system.

The mathematical relation that relates these three parameters is given as

ΔS = qrev / T

Where ΔS = change in the entropy

qrev = heat of the reversible reaction

T = temperature

Question 13. An increase In the enthalpy of the surroundings is the same as the decrease In the enthalpy of the system. Will the temperature of the system and its surroundings be equal if they are in thermal equilibrium?

Answer 13. Yes, if the system, as well as the surroundings, are in thermal equilibrium, their temperatures are equal. Also, an increase in the enthalpy of the surroundings is equal to a decrease in the enthalpy of the system.

Question 14. At the temperature 298 K, the Kp of the reaction is given as N_2O_4 (g) ⇌ $2NO_2$ (g) is 0.98. Predict when the reaction is spontaneous or not.

Answer 14. For the spontaneous reaction,

$\Delta_r G°$ is given as -ve.

$\Delta_r G°$ = – RT ln Kp = – RT ln (0.98)

In (0.98) is – 0.02

$\Delta_r G°$ = – RT × –0.02

Δ will be positive.

Thus, the reaction will be considered as non-spontaneous.

Question 15. The sample for 1.0 mol of the monatomic ideal gas is taken through the cyclic process of expansion as well as compression, as shown in Fig. 6.1. What will be the value for the ΔH of the cycle as the whole?

Answer 15. The change in the internal energy during the cyclic process is zero.

$\Delta U = 0$

When the system returns to the initial state, no work is said to have taken place.

Enthalpy for the steady-state cyclic process is one single value at any given stage, but it is different at the different stages.

$\Delta H = 0$

Question 16. The standard molar entropy for H2O (l) is 70 J K-1 mol-1. Will the standard molar entropy for H2O (s) be more or less as compared to 70 J K-1 mol-1?

Answer 16. The solid form for the H2O is ice. In ice, the molecules are less random as compared with liquid water.

So, the molar entropy of H2O (s) < the molar entropy of H2O (l).

The standard molar entropy for H2O (s) is less compared with 70 J K-1 mol-1.

Question 17. Identify the state functions as well as path functions for the following: enthalpy, entropy, heat, temperature, work, and free energy.

Answer 17. State Functions are Enthalpy, Entropy, Temperature, and Free energy.

Path Functions are Heat and Work.

Question 18. The molar enthalpy for the vaporisation of acetone is less as compared with that of the water. Why?

Answer 18. As the acetone lacks a hydrogen bond, the intermolecular forces are weaker, causing it to boil/evaporate quickly, causing lower the molar enthalpy for the vaporisation. Moreover, as acetone lacks the polar O-H bond, it has a low enthalpy. Water has both the non-polar region as well as a strong hydrogen bond.

Question 19. Which quantity out of the following ΔrG and ΔrG⊖ will be zero at equilibrium?

Answer 19. ΔrG will always be zero.

ΔrG⊖ is zero for K = 1 as ΔrG⊖ = – RT ln K

ΔrG⊖ will be considered non-zero for the other values of K.

Question 20. Predict the change in the internal energy for the isolated system at the constant volume.

Answer 20. There is no energy transfer as the heat or the work in the isolated system,

Thus, w=0 and q=0.

According to the first law of thermodynamics-

ΔU = q + w = 0 + 0 = 0

ΔU = 0

Question 21. Although heat is the path function, heat absorbed by the system under certain specific conditions is independent of the path. What are the conditions for it? Explain.

Answer 21. (i) At the constant volume-

By the first law of thermodynamics:

q = ΔU + (–w)

(–w) = pΔV

q = ΔU + pΔV

ΔV = 0,

As the volume is constant.

$q_v = \Delta U + 0 \Rightarrow q_v = \Delta U$ = change in the internal energy

(ii) At the constant pressure

$q_p = \Delta U + p\Delta V$

However, $\Delta U + p\Delta V = \Delta H$

$q_p = \Delta H$ = change in the enthalpy.

Hence, at the constant volume as well as at the constant pressure, the heat change is the state function as it is equal to the change in the internal energy as well as the change in the enthalpy respectively that are in the state functions.

Question 22. Expansion of the gas in the vacuum is called free expansion. Calculate the work done as well as the change in the internal energy if 1 litre of the ideal gas expands isothermally into a vacuum until the total volume is 5 litres?

Answer 22.

$(-w) = p_{ext}(V_2 - V_1) = 0 \times (5 - 1) = 0$

For the isothermal expansion,

$q = 0$

By the first law of thermodynamics-

$q = \Delta U + (-w)$

$0 = \Delta U + 0$

Hence, $\Delta U = 0$

Question 23. Heat capacity (C_p) is the extensive property; however, the specific heat (c) is the intensive property. What will be the relation between the C_p and c for 1 mol of the water?

Answer 23.

For the water, the heat capacity is = 18 × specific heat

And $C_p = 18 \times c$

Specific heat is

$c = 4.18$ Jg-1 K-1

Heat capacity is

$C_p = 18 \times 4.18$ JK-1 = 75.3 J K-1.

Question 24. The difference between the Cp and Cv can be derived as the empirical relation H = U + pv. Calculate the difference between the Cp and Cv for the ten moles of the ideal gas.

Answer 24.

For the 1 mole of the gas,

$C_p - C_v = R$

For n moles of the gas, the relation is given as

$C_p - C_v = nR = 10 \times 4.184$ J

$C_p - C_v = 41.84$ J.

Question 25. When the combustion of 1g of the graphite produces 20.7 kJ of heat, what will be the molar enthalpy change? Give the significance of the sign also.

Answer 25.

The molar enthalpy change of the graphite = enthalpy change for the 1g carbon × molar mass of the carbon

$\Delta H = -20.7$ kJ g-1 × 12 g mol-1

$\Delta H = -2.48 \times 10^2$ kJ mol-1.

The exothermic nature of the reaction is indicated with the

negative sign for ΔH.

Question 26. The net enthalpy change for the reaction is the amount of energy needed to break all the bonds from the reactant molecules minus the amount of energy needed to form all the bonds in the product molecules. What will be the enthalpy change for the given reaction?

$H_2(g) + Br_2(g) \rightarrow 2HBr(g)$

Given that the Bond energy of the H2, Br2 and HBr is 435 kJ mol-1, 192 kJ mol-1, as well as 368 kJ mol-1, respectively.

Answer 26. $\Delta_r H^\ominus$ = Bond energy for H2 + Bond energy for Br2 − 2 × Bond energy for HBr

$\Delta_r H^\ominus$ = 435 + 192 − (2 × 368) kJ mol-1

$\Delta_r H^\ominus$ = −109 kJ mol-1.

Question 27. What will be the work done of an ideal gas enclosed in the cylinder if it is compressed by the constant external pressure, text in the single step as shown in Fig. Explain it graphically.

Answer 27. Work done on the ideal gas can be calculated based on the area covered by the P-V graph (shaded region) is the actual value of the work don = length × breadth = pext ΔV = AVI (or BVII) × (VI-VII)

Question 28. How will you calculate the work done on the ideal gas in the compression if a change in the pressure Is carried out in infinite steps?

Answer 28. The process or change is said to be reversible when it is carried out in such a way that the process can be reversed at any time by the infinitesimal change.

When pressure is not constant and changes occur in it at the infinite steps (reversible conditions) during the compression from the initial volume, Vi, to the final volume, Vf, work done could be calculated using the pV-plot. The shaded area shows the work done on the gas.

Question 29. Represent the potential energy/enthalpy change in the given processes graphically.

(a) Throwing a stone from the ground to the roof.

(b) ½ H2(g) + ½ Cl2(g) ⇌ HCl(g); ΔrHΘ = –92.32 kJ mol-1

In which of the processes potential energy/enthalpy change is the contributing factor to the spontaneity?

Answer 29. (a) When throwing the stone from the ground to the roof, we must give energy to the stone.

(b) When the heat is produced in the reaction, it indicates that the decrease follows the process in the energy.

Energy increases in (a) part while decreases in (b). As a result, in process (b), the enthalpy change is the factor that contributes to spontaneity.

Question 30. Enthalpy diagram for the particular reaction is given in Fig. Is it possible to decide the reaction's spontaneity from the given diagram? Explain.

Answer 30. No, enthalpy is not the only criterion for determining spontaneity; we must also consider entropy.

COMPETITIVE CORNER

NEET

1. What will be the value of ΔH, if the forward and reverse reactions have the same energy of activation?

(a) ΔH = ΔG = ΔS = 0

(b) ΔS = 0

(c) ΔG = 0

(d) ΔH = 0

Answer: (d)

2. What will be the entropy change (ΔS), when an ideal gas undergoes a change in the pressure from p_i to p_f isothermally?

ANS : NIL

4. What is the molar entropy change for melting of ice at 0°C, if ΔH_f = 1.435 kcal/mol?

(a) 0.526 cal/(mol K)

(b) 5.26 cal/(mol K)

(c) 10.52 cal/(mol K)

(d) 21.04 cal/(mol K)

Answer: (b)

5. What is the function of a catalyst in a chemical reaction?

(a) decrease rate constant of reaction

(b) increases activation energy of reaction

(c) reduces enthalpy of reaction

(d) does not affect the equilibrium constant of reaction

Answer: (d)

6. What will be the work done by 3 moles of an ideal gas when it expands spontaneously in a vacuum?

(a) zero

(b) infinite

(c) 3 joules

(d) 9 joules

Answer: (a)

7. Find the temperature at which the below reaction will be in equilibrium if the enthalpy and entropy change for the reaction is 30 kJ mol^{-1} and 105 J K^{-1} mol^{-1} respectively

$Br_{2(l)} + Cl_{2(g)} \rightarrow 2BrCl_{(g)}$

(a) 273 K

(b) 300 K

(c) 450 K

(d) 285.7 K

Answer: (d)

8. Which is true for the entropy of a spontaneous reaction?

(a) $\Delta S_{(system)} - \Delta S_{(surroundings)} > 0$

(b) $\Delta S_{(system)} + \Delta S_{(surroundings)} > 0$

(c) $\Delta S_{(surroundings)} > 0$ only

(d) $\Delta S_{(system)} > 0$ only

Answer: (b)

9. An ideal gas is expanded isothermally at 300 K from 1 litre to 10 litres. Find the ΔE for this process (R = 2 cal mol^{-1} K^{-1})

(a) 9 L atm

(b) 1381.1 cal

(c) zero

(d) 163.7 cal

Answer: (c)

10. ΔH for the reaction $N_2 + 3H_2 \rightleftharpoons 2NH_3$

(a) ΔE – 2RT

(b) ΔE + 2RT

(c) ΔE – RT

(d) ΔH = RT

Answer: (a)

JEE

Q1: "Heat cannot by itself flow from a body at a lower temperature to a body at a higher temperature" is a statement or consequence of

(a) The second law of thermodynamics

(b) conservation of momentum

(c) conservation of mass

(d) The first law of thermodynamics

Answer: (a) Second law of thermodynamics.

Q2: Which of the following is incorrect regarding the first law of thermodynamics?

(a) It introduces the concept of internal energy

(b) It introduces the concept of entropy

(c) It is not applicable to any cyclic process

(d) It is a restatement of the principle of conservation of energy

Answer: Statements (b) and (c) are incorrect regarding the first law of thermodynamics.

Q3: Which of the following statements is correct for any thermodynamic system?

(a) The internal energy changes in all processes

(b) Internal energy and entropy are state functions

(c) The change in entropy can never be zero

(d) The work done in an adiabatic process is always zero

Answer: (b) Internal energy and entropy are state functions

Q4: Which of the following parameters does not

characterize the thermodynamic state of matter?

(a) temperature

(b) pressure

(c) work

(d) volume

Answer: (c): The work does not characterize the thermodynamic state of matter

Q5: Even a Carnot engine cannot give 100% efficiency because we cannot

(a) prevent radiation

(b) find ideal sources

(c) reach absolute zero temperature

(d) eliminate friction.

Answer: (c): We cannot reach absolute zero temperature

Q6: Which statement is incorrect?

(a) All reversible cycles have the same efficiency

(b) The reversible cycle has more efficiency than an irreversible one

(c) Carnot cycle is a reversible one

(d) Carnot cycle has the maximum efficiency in all cycles

Answer: (a) All reversible cycles do not have the same efficiency.

Q7: A Carnot engine operating between temperatures T_1 and T_2 has efficiency 1/6. When T_2 is lowered by 62 K, its efficiency increases to ⅓. Then T_1 and T_2 are, respectively

(a) 372 K and 310 K

(b) 372 K and 330 K

(c) 330 K and 268 K

(d) 310 K and 248 K

Solution

The efficiency of Carnot engine,

$\eta = 1 - (T_2/T_1)$

$\eta = 1/6$

$T_2/T_1 = 5/6$

$T_1 = 6T_2/5$ ———————(1)

As per the question, when T_2 is lowered by 62 K, then its efficiency becomes 1/3

$1/3 = [1 - (T_2 - 62/T_1)] [T_2 - 62/T_1] = 1 - (1/3)$ (Using equa (1))

$5(T_2 - 62)/6T_2 = 2/3$

$5T_2 - 310 = 4T_2 \Rightarrow T_2 = 310$ K

From equation (1) T1 = (6 x 310)/5 = 372 K

Answer: (a) 372 K and 310 K

Q8: 100 g of water is heated from 30 °C to 50 °C. Ignoring the slight expansion of the water, the change in its internal energy is (specific heat of water is 4184 J $kg^{-1} K^{-1}$)

(a) 4.2 kJ

(b) 8.4 kJ

(c) 84 kJ

(d) 2.1 kJ

Solution

$\Delta Q = ms\Delta T$

Here m = 100 g = 100 x 10^{-3} Kg

S = 4184 J $kg^{-1}K^{-1}$ and ΔT = (50 – 30) = 20 °C

ΔQ = 100 x 10^{-3} x 4184 x 20 = 8.4 x 10^3 J

$\Delta Q = \Delta U + \Delta W$

Change in internal energy

$\Delta U = \Delta Q$ = 8.4 x 10^3 J = 8.4 kJ

Answer: (b) 8.4 kJ

Q9: 200 g water is heated from 40° C to 60 °C. Ignoring the slight expansion of water, the change in its internal energy is close to (Given specific heat of water = 4184 J/kg/K)

(a) 167.4 kJ

(b) 8.4 kJ

(c) 4.2 kJ

(d) 16.7 kJ

Solution

For isobaric process, $\Delta U = Q = ms\Delta T$

Here, m = 200 g = 0.2 Kg, s = 4184 J/Kg/K

$\Delta T = 60\,°C - 40\,°C = 20\,°C$

$\Delta U = (0.2)(4184)(20) = 16736\ J = 16.7\ kJ$

Answer: (d) 16.7 kJ

Q10: The work of 146 kJ is performed in order to compress one-kilo mole of gas adiabatically and in this process the temperature of the gas increases by 7°C. The gas is ($R = 8.3\ J\ mol^{-1}\ K^{-1}$)

(a) monoatomic

(b) diatomic

(c) triatomic

(d) a mixture of monoatomic and diatomic

Solution

According to the first law of thermodynamics

$\Delta Q = \Delta U + \Delta W$

For an adiabatic process, $\Delta Q = 0$

$0 = \Delta U + \Delta W$

$\Delta U = -\Delta W$

$nC_v\Delta T = -\Delta W$

$C_v = -\Delta W/n\Delta T = -[-146 \times 10^3]/[(1 \times 10^3) \times 7] = 20.8$ Jmol^{-1}K^{-1}

For diatomic gas, $C_v = (5/2)R = (5/2) \times 8.3 = 20.8$ Jmol^{-1}K^{-1}

Hence, the gas is diatomic

Answer: (b) diatomic

Q11: From the following statements, concerning ideal gas at any given temperature T, select the correct.

(a) The coefficient of volume expansion at constant pressure is the same for all ideal gases

(b) The average translational kinetic energy per molecule of oxygen gas is 3kT, k being Boltzmann constant

(c) The mean-free path of molecules increases with an increase in the pressure

(d) In a gaseous mixture, the average translational kinetic energy of the molecules of each component is different

Solution

$\gamma = dV/(V_0 \times dT)$ at a constant temperature

$\gamma = 1/V_0 (dV/dT)_p$ since $PV = RT$

$PdV = RdT$ or $(dV/dT) = R/P_0$

Therefore, $\gamma = (1/V_0)(R/P_0) = R/RT_0$

$\gamma = 1/T_0$

$\gamma = 1/273$

Answer: (a) The coefficient of volume expansion at constant pressure is the same for all ideal gases

Q12: Calorie is defined as the amount of heat required to raise the temperature of 1 g of water by 1 °C and it is defined under which of the following conditions?

1. From 14.5 °C to 15.5 °C at 760 mm of Hg
2. From 98.5 °C to 99.5 °C at 760 mm of Hg
3. From 13.5 °C to 14.5 °C at 76 mm of Hg
4. From 3.5 °C to 4.5 °C at 76 mm of Hg

Solution

1 calorie is the amount of heat required to raise the temperature of 1gm of water from 14.5 ^{0}C to 15.5 ^{0}C at

760 mm of Hg

Answer: (a) From 14.5 °C to 15.5 °C at 760 mm of Hg

Q13: The average translational kinetic energy of O2 (molar mass 32) molecules at a particular temperature is 0.048 eV. The translational kinetic energy of N2 (molar mass 28) molecules in eV at the same temperature is

(a) 0.0015

(b) 0.003

(c) 0.048

(d) 0.768

Solution

Average Kinetic Energy = (3/2)KT

It depends on temperature and does not depend on molar mass

For both the gases, average translational kinetic energy will be the same ie., 0.048 eV

Answer: (c) 0.048

Q14: One mole of a monoatomic gas is heated at a constant pressure of 1 atmosphere from 0K to 100K. If the gas constant R=8.32 J/mol K, the change in internal energy of the gas is approximately [1998]

(a) 2.3 J

(b) 46 J

(c) 8.67×10^3 J

(d) 1.25×10^3 J

Solution

$\Delta U = nC_v dT = 1 \times (3R/2) \Delta T$

$\Delta U = (3/2) \times (8.3) \times (100) = 1.25 \times 10^3$ J

Answer: (d) 1.25×10^3 J

Q15: An ideal gas heat engine is operating between 227 °C and 127 °C. It absorbs 10^4 J of heat at a higher temperature. The amount of heat converted into work is

(a) 2000 J

(b) 4000 J

(c) 8000 J

(d) 5600 J

Solution

$\eta = 1 - (T_2/T_1)$

$\eta = 1 - (127 + 273)/(227 + 273) = 1 - (400/500) = 1/5$

$W = \eta Q_1 = 1/5 \times 10^4 = 2000$ J

Answer: (a) 2000 J

EQUILIBRIUM

Question 1. We know about the relationship between Kc and Kp is $Kp = Kc(RT)^{\Delta n}$

What will be the value of Δn for the reaction

$NH_4Cl\ (s) \rightarrow NH_3\ (g) + HCl\ (g)$

(i) 1

(ii) 0.5

(iii) 1.5

(iv) 2

Answer 1: Option (iv) is the correct answer.

Question 2. In the case of the reaction $H_2(g) + I_2(g) \rightarrow 2HI\ (g)$, the standard free energy is given as

$\Delta G > 0$.

The equilibrium constant (K) will be ___.

(i) K = 0

(ii) K > 1

(iii) K = 1

(iv) K < 1

Answer 2: Option (iv) is the correct answer.

Question 3. Which of the following is not the general feature of equilibria involving physical processes?

(i) Equilibrium is possible only in the closed system at the given temperature.

(ii) All measurable properties of the given system remain constant.

(iii) All the physical processes stop at equilibrium.

(iv) The opposing process occurs at the same rate, and there is the dynamic, however stable condition.

Answer 3: Option (iii) is the correct answer.

Question 4. PCl_5, PCl_3, and Cl_2 are at equilibrium at 500K in the closed container, and their concentrations are given as 0.8×10^{-3} mol L^{-1}, 1.2×10^{-3} mol L^{-1}, and 1.2×10^{-3} mol L^{-1} respectively. The value of K_c for the given reaction

$PCl_5 (g) \rightleftharpoons PCl_3 (g) + Cl_2 (g)$

will be,

(i) 1.8×10^{3} mol L^{-1}

(ii) 1.8×10^{-3}

(iii) 1.8×10^{-3} L moL^{-1}

(iv) 0.55×10^{4}

Answer 4: Option (ii) is the correct answer.

Question 5. If hydrochloric acid is added to the cobalt nitrate solution at room temperature, the following reaction will occur, and the reaction mixture will become blue. On cooling the mixture, it becomes pink. Based on the given information, mark the right answers.

$[Co(H_2O)_6]^{3+} (aq) + 4Cl^- (aq) \rightleftharpoons [CoCl_4]^{2-} (aq) + 6H_2O (l)$

(pink) (blue)

(i) ΔH > 0 for the reaction

(ii) ΔH < 0 for the reaction

(iii) ΔH = 0 for the reaction

(iv) The sign of ΔH can't be predicted based on the given information.

Answer 5; Option (i) is the correct answer.

Question 6. The pH of the neutral water at 25°C is 7.0. When the temperature increases, the ionisation of water increases, but the concentration of H+ ions as well as OH– ions are the same. What would be the pH of pure water at 60°C?

(i) Equal to 7.0

(ii) Greater than 7.0

(iii) Less than 7.0

(iv) Equal to zero

Answer 6: Option (iii) is the correct answer.

Question 7. Ka, 2Ka, and 3Ka are the respective ionization constants for the following given reactions.

$H_2S \rightleftharpoons H^+ + HS^-$

$HS^- \rightleftharpoons H^+ + S^{2-}$

$H_2S \rightleftharpoons 2H^+ + S^{2-}$

The correct relationship in between Ka1, Ka2 and Ka3 will be

(i) Ka3 = Ka1 × Ka2

(ii) Ka3 = Ka1 + Ka2

(iii) Ka3 = Ka1 – Ka2

(iv) Ka3 = Ka1 /Ka2

Answer 7: Option (i) is the correct answer.

Question 8. The acidity of the compound BF3 could be explained based on which among the following concepts?

(i) Arrhenius's concept

(ii) Bronsted Lowry's concept

(iii) Lewis's concept

(iv) Bronsted Lowry as well as Lewis's concept.

Answer 8: Option (iii) is the correct answer.

Question 9. Which among the following will produce the buffer solution if mixed in equal volumes?

(i) 0.1 mol dm–3 NH4OH as well as 0.1 mol dm–3 HCl

(ii) 0.05 mol dm–3 NH4OH as well as 0.1 mol dm–3 HCl

(iii) 0.1 mol dm–3 NH4OH as well as 0.05 mol dm–3 HCl

(iv) 0.1 mol dm–3 CH4COONa as well as 0.1 mol dm–3 NaOH

Answer 9. Option (iii) is the correct answer.

Question 10. Which among the following solvents in silver chloride is most soluble?

(i) 0.1 mol dm–3 AgNO3 solution

(ii) 0.1 mol dm–3 HCl solution

(iii) H2O

(iv) Aqueous ammonia

Answer 10. Option (iv) is the correct answer.

Question 11. The ionization for hydrochloric in the water has given below:

$HCl(aq) + H2O (l) \rightleftharpoons H3O+(aq) + Cl-(aq)$

Label the two conjugate acid-base pairs in this ionization.

Answer 11: The two required conjugate acid-base pairs in the ionization of HCl are (HCl–Cl–), where HCl is the conjugate acid as well as Cl– is the conjugate base. In the same way, the second pair is (H2O–, H3O+), where H2O is the conjugate base and H3O+ is the conjugate acid.

Question 12. The conjugate acid of the weak base is always stronger. What would be the decreasing order of the basic strength of the following conjugate bases?

OH–, RO–, CH3COO–, Cl–

Answer 12: The conjugate base of the strong acid is weak thud the decreasing order of the basic strength would be;

RO– > OH– > CH3COO– > Cl–

Question 13. The aqueous solution of the sugar does not conduct electricity. But, when sodium chloride is added to water, it conducts electricity. How would you explain the statement based on the ionisation, as well as how is it affected by the concentration of the sodium chloride?

Answer 13; The aqueous solution of sugar does not conduct electricity as they exist as the molecule in the water. They don't have free ions to conduct electricity; however, in the case of NaCl, free ions of Na+ as well as Cl– are present to conduct the electricity. Conductance depends on the no. of the ions present in the given solution. More will the given no. of ions of NaCl in water more will be the conductivity.

Question 14. The reaction between ammonia and boron trifluoride is given below as follows:

NH3 + BF3 → H3N: BF3

Identify the acid and base in the given reaction. Which theory gives an explanation to it? What is the hybridisation of B as w N in the given reactants?

Answer 14: NH3 is the Lewis base, while BF3 is the Lewis acid. Lewis's electronic theory of acids, as well as bases, explains it.

The hybridisation state of the nitrogen in NH3 is sp3 hybridised, as well as Boron in BF3, which is sp2 hybridised.

Question 15. The compound BF3 does not have a proton; however, it still acts as an acid and reacts with NH3. Why is it so? What type of bond can be formed between the two?

Answer 15: According to the Lewis concept, e- deficient species are known as lewis acid. Thus BF3 will act as the lewis acid while NH3 (N=1S2 2S2 2p3) has the lone pair; hence it will act as the lewis base, and it will donate the lone pair to the empty p-orbital of the Boron through the coordinate bond to form an adduct.

Question 16. Based on the given equation pH = – log [H+], the pH of the given 10-8 mol dm-3 solution of HCl shall be 8. But, it is observed to be less than 7.0. Justify the reason.

Answer 16: The solution is given very dilute, and we know that HCl reacts with water to give hydronium ions. The decrease in the pH could be observed as the result of the large concentration of H+. Hydronium ion concentration also needs to be considered here.

Then, the total pH would be;

[H3O+] = 10-8 + 10-7 M = 7

Thus, the solution will be acidic.

Question 17. The ionisation constant of the weak base MOH is given by the expression;

$K_b = [M^+][OH^-]/[MOH^-]$

Values of the ionisation constant for some weak bases at particular temperatures give below:

Base: Di-methylamine, Urea, Pyridine, and Ammonia

K_b: 5.4×10^{-4}, 1.3×10^{-14}, 1.77×10^{-9}, 1.77×10^{-5}

Arrange the following bases in the decreasing order of the extent of their ionisation at equilibrium. Which among the above base is the strongest?

Answer 17: The decreasing order of the bases based on the ionisation constant at equilibrium would be;-

Di-methylamine > Ammonia > Pyridine > Urea

The strongest base would be Di-methyl amine because its pKb value is 3.29, and we already know that the less the pKb value, the strong is the base.

Question 18. Arrange the following compounds in increasing order for pH

KNO_3 (aq), CH_3COONa (aq), NH_4Cl (aq), $C_6H_5COONH_4$ (aq)

Answer 18: The increasing order of the pH would be;

$CH_3COONa < KNO_3 < C_6H_5COONH_4 < NH_4Cl$

CH_3COONa is the salt of a weak acid (CH_3COOH) and strong base (NaOH)

KNO_3 is the salt of strong acid (HNO_3)-strong base (KOH)

$C_6H_5COONH_4$ is the salt of a weak acid (benzoic acid) and weak base (NH_4OH)

NH_4Cl is the salt of a strong acid (HCl) and weak base (NH_4OH)

Question 19. Conjugate acid of the weak base is always

stronger. The decreasing order of the basic strength for the following conjugate bases will be?

OH-, RO-, CH3COO-, CI-

Answer 19: Conjugate acids of the given bases are H2O, ROH, CH3COOH, and HCl.

Their acidic strength has the order.

HCl > CH3COOH > H2O > ROH.

Therefore, their conjugate bases would have strength in the order.

RO- > OH- > CH3COO- > CI-

Question 20. The value of Kc for the given reaction 2HI (g) ⇌ H2 (g) + I2 (g) is 1 × 10-4

At the given time, the composition of the reaction mixture is given as follows;

[HI] = 2 × 10-5 mol, [H2] = 1 × 10-5 mol as well as [I2] = 1 × 10-5 mol.

In which direction would the reaction proceed?

Answer 20: Given as;

Kc = 1×10-4

Kc = [H2][I2]/[HI]2

Qc expresses the relative ratio of the products to reactants at the given instant.

Qc = [H2][I2]/[HI]2

= (1×10-5)(1×10-5)/(2×10-4)

Qc = 1/4 = 0.25

Where; Qc > Kc, the reaction would proceed in the reverse

direction.

Question 21. When 0.561 g of KOH is dissolved in water to obtain 200 mL of solution at 298 K. Calculate the concentrations for the potassium, hydrogen and hydroxyl ions. What is the pH?

Answer 21. [KOH(aq)] = 0.561 / (1/5)g/L

= 2.805 g/L

= 2.805 x (1/56.11)

= 0.05M

$KOH(aq) \rightarrow 10^{-13}(aq) + OH^-(aq)$

$[OH^-] = 0.05M = [10^{-13}] [H^+][OH^-] = K_w [H^+] = K_w/[OH^-] [H^+] = 10^{-14}/0.05 [H^+] = 2 \times 10^{-13} M$

$pH = -\log[H^+]$

$pH = -\log[2 \times 10^{-13}]$

pH = 12.70

Question 22. The pH of the given 0.08 mol dm^{-3} HOCl solution is 2.85. Calculate the ionisation constant.

Answer 22: pH of the given HOCl = 2.85

However, $-pH = \log[H^+]$

So, $-2.85 = \log[H^+]$

$-3.15 = \log[H^+] [H^+] = 1.413 \times 10^{-3}$

For the given weak monobasic acid $[H^+] = (K_a \times C)^{1/2}$

$K_a = [H^+]^2 / C$

$K_a = (1.413 \times 10^{-3})^2 / 0.08$

$K_a = 24.957 \times 10^{-6}$

$K_a = 2.4957 \times 10^{-5}$

Question 23. The sparingly soluble salt gets precipitated only if the product of the concentration of the ions in the given solution (Qsp) becomes greater than the solubility product. When the solubility of BaSO4 in water is 8×10^{-4} mol dm^{-3}, calculate the solubility in 0.01 mol dm^{-3} of H2SO4.

Answer 23: Given that the standard solubility of BaSO4 in water is 8×10^{-4} g/L

The equation for the disassociation of BaSO4 would be-

$BaSO_4 \rightleftharpoons Ba^{2+} + SO_4^{2-}$

(S' is the solubility of the Ba2+ in 0.01 of HCl)

S <<< 0.01, thus it could be neglected

We already know that $K_{sp} = S^2$

$K_{sp} = (8 \times 10^{-8})^2$

$= 64 \times 10^{-8}$

Then, $K_{sp} = (S')(0.01)$

$S' = 64.8 \times 10^{-8}/0.01 = 6.4 \times 10^{-5}$

Thus, the solubility of BaSO4 in the given 0.01 mol dm^{-3} of H2SO4 is 6.4×10^{-5}

Question 24. Calculate the hydrogen ion concentration for the following biological fluids which have pH as given below:

(I) For Human saliva, 6.4

(II) For Human stomach fluid, 1.2

(III) For Human muscle fluid, 6.83

(IV) For Human blood, 7.38.

Answer 24. (I) For Human saliva, 6.4

pH = 6.4

$6.4 = -\log[H+]$ $[H+] = 3.98 \times 10^{-7}$

(II) For Human stomach fluid, 1.2

pH = 1.2

$1.2 = -\log[H+]$

∴ $[H+] = 0.063$

(III) For Human muscle fluid, 6.83

pH = 6.83

$pH = -\log[H+]$

∴ $6.83 = -\log[H+]$ $[H+] = 1.48 \times 10^{-7}$ M

(IV) For Human blood, 7.38

$pH = 7.38 = -\log[H+]$

∴ $[H+] = 4.17 \times 10^{-8}$ M

Question 25. On the basis of the given equation pH = − log [H+], the pH of the given 10-8 mol dm-3 solution for HCI should be 8. But, it is observed as less than 7.0. Justify the reason.

Answer 25: The water concentration could not be neglected as the solution is very dilute.

$[H3O+] = 10^{-8} + 10^{-7}$ M

$[H3O+] = 10^{-8}(1 + 10)$

$[H3O+] = 11 \times 10^{-8}$ M

$pH = -\log[H3O+]$

$pH = -\log 11 \times 10^{-8}$ M

pH = 8 – log 11

pH = 8 – 1.04

pH = 6.96

pH would be less than 7.0.

Question 26. The pH of the given 0.1M solution of cyanic acid (HCNO) is 2.34. Calculate the ionisation constant of the given acid and the degree of ionisation in the solution.

Answer 26. c = 0.1 M

pH = 2.34

-log [H+] = pH

-log [H+] = 2.34

[H+] = 4.5 × 10^{-3}

And

[H+] = cα

4.5 × 10^{-3} = 0.1 × α

α = 0.1/(4.5 × 10^{-3})

α = 0.045

Ka = c$α^2$

Ka = 0.1 × $(0.045)^2$

Ka = 0.0002025

Ka = 2.025 × 10^{-4}

Question 27. The value of the given Kc for the given reaction 2HI (g) ⇌ H2 (g) + I2 (g) is 1 × 10^{-4}. At the given time, the composition of the reaction mixture is [HI] = 2 × 10^{-5} mol, [I2] = 1 × 10^{-5} mol. Determine the direction of how the reaction would proceed?

Answer 27: At the given time, the reaction quotient Q for the reaction would be given by the expressions.

$Q = [H_2][I_2] / [HI]$

$Q = 1 \times 10^{-5} \times 1 \times 10^{-5} / (2 \times 10^{-5})^2$

$Q = 1/4$

$Q = 0.25$

$Q = 2.5 \times 10^{-1}$

Since the value of the given reaction quotient is greater than the value of Kc, i.e. 1×10^{-4}, the reaction would proceed in the reverse direction.

Question 28. The pH of the given solution of the strong acid is 5.0. What would be the pH of the given solution obtained after diluting the given solution 100 times?

Answer 28: pH = 5 means $[H^+] = 10^{-5}$

On diluting 100 times, we get,

$[H^+] = 10^{-5} / 100 = 10^{-7}$

On calculating the pH of the given equation, we get,

pH = $-\log[H^+]$, pH value comes out as 7, that is impossible.

Therefore, the total H^+ ion concentration = H^+ ions for the given acid H^+ ion from water

$[H^+] = 10^{-7} + 10^{-7}$ M

$[H^+] = 2 \times 10^{-7}$

pH = 7 − 0.3010

pH = 6.699

Question 29. Predict whether the solutions for the following salts are neutral, acidic or basic: NaCl, KBr, NaCN, NH4NO3, NaNO2 and KF.

Answer 29.

KBr
KBr + H2O ↔ KOH (Strong base)+ HBr (Strong acid) Therefore, it is a neutral solution.

NH4NO3
NH4NO3 + H2O ↔ NH4OH(Weak base) + HNO2 (Strong acid)

Therefore, it is an acidic solution.

KF
KF + H2O ↔ KOH (Strong base) + HF (weak acid)

Therefore, it is a basic solution.

NaNO2
NaNO2 + H2O ↔ NH4OH(Strong base) + HNO2(Weak acid)

Therefore, it is a basic solution.

NaCN
NaCN + H2O ↔ HCN (Weak acid) + NaOH (Strong base)

Therefore, it is a basic solution.

NaCl
NaCl + H2O ↔ NaOH (Strong base) + HCL (Strong acid)

Therefore, it is a neutral solution.

Question 30. Calculate the pH of the given solution formed by mixing equal volumes of the two solutions A and B of the strong acid having pH = 6 and pH = 4, respectively.

Answer 30: The pH of the given solution A = 6,

Thus, the concentration of the [H+] ion in solution A=10-6

mol / L

The pH of the given solution B = 4

Thus, the concentration of the [H+] ion in solution B = 10^{-4} mol / L.

On mixing, one litre of each given solution, the total volume = 1L + 1L = 2L.

Amount of the given H+ ions in 1L of the solution A = concentration × Volume (V) =

Amount of the given H+ ions in 1L of the solution A = 10^{-6} × 1 = 10^{-6}

Amount of the given H+ ions in 1L of the solution B = 10^{-4} × 1 = 10^{-4}

The total amount of the given H+ ions in the solution formed by mixing solutions A and B is (10^{-5} mol + 10^{-4} mol)

This amount is present in the given 2L solution,

So, Total [H+] = 10^{-4} (1 + 1.01) / 2 mol / L

Total [H+] = 1.01 × 10^{-4} / 2 mol / L

Total [H+] = 0.5 × 10^{-4} mol / L

Total [H+] = 5 × 10^{-5} mol / L

pH = – log [H+]

pH = – log [5 × 10^{-5}]

pH = – [log 5 – 5 log 10]

pH = – log 5 + 5

pH = 5 – log 5

pH = 5 – 0.6990

pH = 4.3010

pH = 4.3

Therefore, the pH will be 4.3.

Question 31. The solubility product of the given compound Al(OH)3 is given as 2.7 x 10-11. Calculate the solubility in g / L and also find out the pH of the given solution. (Atomic mass of Al = 27 u).

Answer 31: Assume S be the solubility of Al(OH)3

$K_{sp} = [Al^{3+}][OH^-]^3$

$K_{sp} = (S)(3S)^3$

$K_{sp} = 27S^4$

$S^4 = K_{sp} / 27$

$S^4 = 27 \times 10^{-11} / 27 \times 10$

$S^4 = 1 \times 10^{-12}$ mol / L.

$S = 1 \times 10^{-3}$ mol / L.

(i) Solubility for Al(OH)3: The molar mass of the given compound Al(OH)3 is 78g.

Hence, the solubility of the given compound Al(OH)3 in g / L = $1 \times 10^3 \times 78$ g / L

Solubility of the given compound Al(OH)3 in g / L = 78×10^{-3} g / L

Solubility of the given compound Al(OH)3 in g / L = 7.8×10^{-2} g / L

(ii) pH of the given solution: $S = 1 \times 10^{-3}$ mol / L

$[OH] = 3S = 3 \times 1 \times 10^{-3} = 3 \times 10^{-3}$

pOH = 3 − log 3

pH = 14 − pOH = 11 + log 3 = 11.4771.

Question 32. The first ionization constant for H2S is 9.1 × 10−8. Calculate the concentration of HS− ion in the 0.1M solution. How would this concentration be affected when the solution is 0.1M in HCl also? If the second dissociation constant for H2S is 1.2 × 10−13, calculate the required concentration of S2− under both conditions.

Answer 32. To calculate for [HS−]

To find [HS−]:

Case 1 − HCl is absent.

Then,

Ka = ([H+][HS−])/[H2S] = 9.1×10-8 (given)

Thus, x2/(0.1-x) = 9.1×10-8

However, 0.1-x is approximately equal to 0.1. Putting the value in the given equation:

x2/0.1 = 9.1×10-8

x2 = 9.1×10-9

x = 9.54× 10-5 M

So, the concentration of HS− is 9.54×10-5 M.

Case 2 − HCl is present

Then,

Ka = ([H+][HS−])/[H2S] = (y× (0.1+y))/(0.1-y) = 9.1×10-8 (given)

However, (0.1 + y) and (0.1 − y) can be approximated to 0.1.

9.1× 10-8 = (0.1*y)/0.1

Thus, y = [HS–] = 9.1×10^{-8} M

To calculate for [S2-]:

Case 1 – HCl is absent.

The dissociation of HS– is given by the following equation below:

HS– $\rightleftharpoons$ H+ + S2-[HS–] = 9.1×10^{-5} M

[H+] = 9.54×10^{-5} M

Ka = ([H+][S2-])/[HS–] = 1.2×10^{-13} (given)

Ka = (9.1×10^{-5} × [S2-])/9.1×10^{-5}

Thus, [S2-] = 1.2×10^{-13} M

Case 2 – HCl is present

[HS–] = 9.1×10^{-8} M [H+] = 0.1 M

Ka = 1.2×10^{-13} M

= (0.1×[S2-])/ 9.1×10^{-8}

Hence, [S2-] = 1.092×10^{-19} M

COMPETITIE CORNER

NEET

1. Find the pH of a solution when 0.01 M HCl and 0.1 M NaOH are mixed in equal volumes

(a) 12.65

(b) 1.04

(c) 7.0

(d) 2.0

Answer: (a)

2. Which of the following aqueous solution will be the best conductor of electricity?

(a) NH_3

(b) CH_3COOH

(c) HCl

(d) $C_6H_{12}O_6$

Answer: (c)

3. In 0.10 M aqueous solution of pyridine (C_5H_5N), find the percentage of pyridine that forms pyridinium ion ($C_5H_5N^+H$) (K_b for C_5H_5N = 1.7 x 10^{-9})

(a) 1.6%

(b) 0.77%

(c) 0.0060%

(d) 0.013%

Answer: (d)

4. Find the equilibrium constant of the reaction

If the equilibrium constant for the following reactions are given

(a) K_2K_3/K_1

(b) $K_1K_3^3/K_2$

(c) $K_2K_3^3/K_1$

(d) $K_2^3K_3/K_1$

Answer: (c)

5. Highest pH will be recorded for which of the following solutions if they are equimolar

(a) $AlCl_3$

(b) $BaCl_2$

(c) $BeCl_2$

(d) LiCl

Answer: (b)

6. The equilibrium constant is 278 for the reaction

at the same temperature, what will be the equilibrium constant for the following reaction?

(a) 6×10^{-2}

(b) 1.8×10^{-3}

(c) 1.3×10^{-5}

(d) 3.6×10^{-5}

Answer: (a)

7. What will be the pH of a buffer solution having an equal concentration of B^- and HB ($K_b = 10^{-10}$ for B^-)

(a) 7

(b) 4

(c) 10

(d) 6

Answer: (b)

8. Find the increase in equilibrium concentration of Fe^{3+} ions if OH^- ions concentration decreases to 1/4th in the following reaction

(a) 8 times

(b) 16 times

(c) 4 times

(d) 64 times

Answer: (d)

9. On increasing the concentration of reactants in a reversible reaction, then equilibrium constant will

(a) depend on the concentration

(b) increase

(c) unchanged

(d) decrease

Answer: (c)

10. Find the conjugate acid of NH_2^-

(a) NH_3

(b) NH_4OH

(c) NH_4^+

(d) NH_2^-

Answer: (a)

JEE

Q1: "Heat cannot by itself flow from a body at a lower temperature to a body at a higher temperature" is a statement or consequence of

(a) The second law of thermodynamics

(b) conservation of momentum

(c) conservation of mass

(d) The first law of thermodynamics

Answer: (a) Second law of thermodynamics.

Q2: Which of the following is incorrect regarding the first law of thermodynamics?

(a) It introduces the concept of internal energy

(b) It introduces the concept of entropy

(c) It is not applicable to any cyclic process

(d) It is a restatement of the principle of conservation of energy

Answer: Statements (b) and (c) are incorrect regarding the first law of thermodynamics.

Q3: Which of the following statements is correct for any thermodynamic system?

(a) The internal energy changes in all processes

(b) Internal energy and entropy are state functions

(c) The change in entropy can never be zero

(d) The work done in an adiabatic process is always zero

Answer: (b) Internal energy and entropy are state functions

Q4: Which of the following parameters does not

characterize the thermodynamic state of matter?

(a) temperature

(b) pressure

(c) work

(d) volume

Answer: (c): The work does not characterize the thermodynamic state of matter

Q5: Even a Carnot engine cannot give 100% efficiency because we cannot

(a) prevent radiation

(b) find ideal sources

(c) reach absolute zero temperature

(d) eliminate friction.

Answer: (c): We cannot reach absolute zero temperature

Q6: Which statement is incorrect?

(a) All reversible cycles have the same efficiency

(b) The reversible cycle has more efficiency than an irreversible one

(c) Carnot cycle is a reversible one

(d) Carnot cycle has the maximum efficiency in all cycles

Answer: (a) All reversible cycles do not have the same efficiency.

Q7: A Carnot engine operating between temperatures T_1 and T_2 has efficiency 1/6. When T_2 is lowered by 62 K, its efficiency increases to ⅓. Then T_1 and T_2 are, respectively

(a) 372 K and 310 K

(b) 372 K and 330 K

(c) 330 K and 268 K

(d) 310 K and 248 K

Solution

The efficiency of Carnot engine,

$\eta = 1 - (T_2/T_1)$

$\eta = 1/6$

$T_2/T_1 = 5/6$

$T_1 = 6T_2/5$ ————(1)

As per the question, when T_2 is lowered by 62 K, then its efficiency becomes 1/3

$1/3 = [1 - (T_2 - 62/T_1)] [T_2 - 62/T_1] = 1 - (1/3)$ (Using equa (1))

$5(T_2 - 62)/6T_2 = 2/3$

$5T_2 - 310 = 4T_2 \Rightarrow T_2 = 310\ K$

From equation (1) $T1 = (6 \times 310)/5 = 372\ K$

Answer: (a) 372 K and 310 K

Q8: 100 g of water is heated from 30 °C to 50 °C. Ignoring the slight expansion of the water, the change in its internal energy is (specific heat of water is 4184 J $kg^{-1} K^{-1}$)

(a) 4.2 kJ

(b) 8.4 kJ

(c) 84 kJ

(d) 2.1 kJ

Solution

$\Delta Q = ms\Delta T$

Here m = 100 g = 100 x 10^{-3} Kg

S = 4184 J $kg^{-1}K^{-1}$ and ΔT = (50 – 30) = 20 °C

ΔQ = 100 x 10^{-3} x 4184 x 20 = 8.4 x 10^3 J

$\Delta Q = \Delta U + \Delta W$

Change in internal energy

$\Delta U = \Delta Q$ = 8.4 x 10^3 J = 8.4 kJ

Answer: (b) 8.4 kJ

Q9: 200 g water is heated from 40° C to 60 °C. Ignoring the slight expansion of water, the change in its internal energy is close to (Given specific heat of water = 4184 J/kg/K)

(a) 167.4 kJ

(b) 8.4 kJ

(c) 4.2 kJ

(d) 16.7 kJ

Solution

For isobaric process, ΔU = Q = msΔT

Here, m = 200 g = 0.2 Kg, s = 4184 J/Kg/K

ΔT = 60 °C − 40 °C = 20 °C

ΔU = (0.2)(4184)(20) = 16736 J = 16.7 kJ

Answer: (d) 16.7 kJ

Q10: The work of 146 kJ is performed in order to compress one-kilo mole of gas adiabatically and in this process the temperature of the gas increases by 7°C. The gas is (R = 8.3 J mol^{-1} K^{-1})

(a) monoatomic

(b) diatomic

(c) triatomic

(d) a mixture of monoatomic and diatomic

Solution

According to the first law of thermodynamics

ΔQ = ΔU + ΔW

For an adiabatic process, ΔQ = 0

$0 = \Delta U + \Delta W$

$\Delta U = -\Delta W$

$nC_v \Delta T = -\Delta W$

$C_v = -\Delta W / n\Delta T = -[-146 \times 10^3]/[(1 \times 10^3) \times 7] = 20.8$ $Jmol^{-1}K^{-1}$

For diatomic gas, $C_v = (5/2)R = (5/2) \times 8.3 = 20.8$ $Jmol^{-1}K^{-1}$

Hence, the gas is diatomic

Answer: (b) diatomic

Q11: From the following statements, concerning ideal gas at any given temperature T, select the correct.

(a) The coefficient of volume expansion at constant pressure is the same for all ideal gases

(b) The average translational kinetic energy per molecule of oxygen gas is 3kT, k being Boltzmann constant

(c) The mean-free path of molecules increases with an increase in the pressure

(d) In a gaseous mixture, the average translational kinetic energy of the molecules of each component is different

Solution

$\gamma = dV/(V_0 \times dT)$ at a constant temperature

$\gamma = 1/V_0 (dV/dT)_p$ since $PV = RT$

$PdV = RdT$ or $(dV/dT) = R/P_0$

Therefore, $\gamma = (1/V_0)(R/P_0) = R/RT_0$

$\gamma = 1/T_0$

$\gamma = 1/273$

Answer: (a) The coefficient of volume expansion at constant pressure is the same for all ideal gases

Q12: Calorie is defined as the amount of heat required to raise the temperature of 1 g of water by 1 °C and it is defined under which of the following conditions?

1. From 14.5 °C to 15.5 °C at 760 mm of Hg
2. From 98.5 °C to 99.5 °C at 760 mm of Hg
3. From 13.5 °C to 14.5 °C at 76 mm of Hg
4. From 3.5 °C to 4.5 °C at 76 mm of Hg

Solution

1 calorie is the amount of heat required to raise the temperature of 1gm of water from 14.5 ^{0}C to 15.5 ^{0}C at

760 mm of Hg

Answer: (a) From 14.5 °C to 15.5 °C at 760 mm of Hg

Q13: The average translational kinetic energy of O2 (molar mass 32) molecules at a particular temperature is 0.048 eV. The translational kinetic energy of N2 (molar mass 28) molecules in eV at the same temperature is

(a) 0.0015

(b) 0.003

(c) 0.048

(d) 0.768

Solution

Average Kinetic Energy = (3/2)KT

It depends on temperature and does not depend on molar mass

For both the gases, average translational kinetic energy will be the same ie.,0.048 eV

Answer: (c) 0.048

Q14: One mole of a monoatomic gas is heated at a constant pressure of 1 atmosphere from 0K to 100K. If the gas constant R=8.32 J/mol K, the change in internal energy of the gas is approximately [1998]

(a) 2.3 J

(b) 46 J

(c) 8.67×10^3 J

(d) 1.25×10^3 J

Solution

$\Delta U = nC_v dT = 1 \times (3R/2) \Delta T$

$\Delta U = (3/2) \times (8.3) \times (100) = 1.25 \times 10^3$ J

Answer: (d) 1.25×10^3 J

Q15: An ideal gas heat engine is operating between 227 °C and 127 °C. It absorbs 10^4 J of heat at a higher temperature. The amount of heat converted into work is

(a) 2000 J

(b) 4000 J

(c) 8000 J

(d) 5600 J

Solution

$\eta = 1 - (T_2/T_1)$

$\eta = 1 - (127 + 273)/(227 + 273) = 1 - (400/500) = 1/5$

$W = \eta Q_1 = 1/5 \times 10^4 = 2000$ J

Answer: (a) 2000 J

REDOX REACTION

Question 1: Which of the following statement(s) is/are not true about the following decomposition reaction?

$2KClO_3 \rightarrow 2KCl + 3O_2$

(i) Potassium is undergoing oxidation

(ii) Chlorine is undergoing oxidation

(iii) Oxygen is reduced

(iv) None of the species are undergoing oxidation or reduction

Answer 1: (i) and (iv)

Explanation: The following statements are incorrect because it is clear from the reaction that oxygen is being oxidised while potassium remains in the same oxidation state.

Question 2: Identify disproportionation reaction

(i) $CH_4 + 2O_2 \rightarrow CO_2 + 2H_2O$

(ii) $CH_4 + 4Cl_2 \rightarrow CCl_4 + 4HCl$

(iii) $2F_2 + 2OH^- \rightarrow 2F^- + OF_2 + H_2O$

(iv) $2NO_2 + 2OH^- \rightarrow NO_2^- + NO_3^- + H_2O$

Answer 2: (iv) $2NO_2 + 2OH^- \rightarrow NO_2^- + NO_3^- + H_2O$

Explanation: Oxidation number of nitrogen decreases by 1 from NO_2 to NO_2^- and increase by +1 from NO_2 to NO_3^-.

Question 3: Which of the following arrangements represents

an increasing oxidation number of the central atom?

(i) CrO_2^-, ClO_3^-, CrO_2^{-4}, MnO^{-4}

(ii) ClO_3^-, CrO_2^{-4}, MnO^{-4}, CrO_2^-

(iii) CrO_2^-, ClO_3^-, MnO^{-4}, CrO_2^{-4}

(iv) CrO_2^{-4}, MnO^{-4}, CrO_2^-, ClO_3^-,

Answer 3: (i) CrO_2^-, ClO_3^-, CrO_2^{-4}, MnO^{-4}

Explanation: As the central element's oxidation number grows as +3, +5, +6, and +7, respectively, the above arrangement depicts the rising oxidation number of the central atom.

Question 4: The oxidation number of an element in a compound is evaluated on the basis of certain rules. Which of the following rules is not correct in this respect?

(i) The oxidation number of hydrogen is always +1.

(ii) The algebraic sum of all the oxidation numbers in a compound is zero.

(iii) An element in the free or uncombined state bears oxidation number zero.

(iv) In all its compounds, the oxidation number of fluorine is -1.

Answer 4: (i)

Explanation: As hydrogen picks up a negative charge when it is with its companion, ionic hydrides have hydrogen in the -1 oxidation state.

Question 5: Using the standard electrode potential, find out the pair between which redox reaction is not feasible.

E values: Fe^{3+}/Fe^{2+} = 0.77; I_2/I^- = + 0.54;

Cu^{2+}/Cu = 0.34; Ag^+/Ag = + 0.80 V

(i) Fe^{3+} and I^-

(ii) Ag^+ and Cu

(iii) Fe^{3+} and Cu

(iv) Ag and Fe^{3+}

Answer 5: (iv) Ag and Fe^{3+}

Explanation: To make the reaction feasible, $E^0$the cell needs to be positive for the pair Ag and Fe^{3+}, but it is negative. Hence the reaction is not feasible.

Question 6: The more positive the value of E, the greater the tendency of the species to get reduced. Using the standard electrode potential of redox couples given below find out which of the following is the strongest oxidising agent.

E values: Fe^{3+}/Fe^{2+} = + 0.77; $I_2(s)/I^-$ = +0.54;

Cu^{2+}/Cu = + 0.34; Ag^+/Ag = + 0.80 V

(i) Fe^{3+}

(ii) $I_2(s)$

(iii) Cu^{2+}

(iv) Ag^+

Answer 6: (iv) Ag^+

Explanation: The reduction potential of Ag^+/Ag is the highest.

Question 7: What is an oxidation reaction?

Answer 7: Oxidation reactions are those which involve either the addition of oxygen or the removal of hydrogen.

Question 8: What is the reduction reaction?

Answer 8: Reduction reactions are those in which oxygen is either removed, or hydrogen is added.

Question 9: What are the most essential conditions that must be satisfied in a redox reaction?

Answer 9: It shouldn't interfere with the theory of electron conservation. The total amount of electrons lost must balance the number of electrons gained by the oxidising agent.

Question 10: What happens to the oxidation number of an element in oxidation?

Answer 10: During oxidation, the oxidation number of the element increases. It is oxidised if the oxidation number increases from 0 to +1.

Question 11: Name the different types of redox reactions.

Answer 11: The different types of redox reactions are:

Combination reactions
Decomposition reactions
Displacement reactions
Disproportionate reactions

Question 12: All decomposition reactions are not redox reactions. Give justification. .

Answer 12: Calcium carbonate decomposition is not a redox reaction. This is true because there must be at least one elemental substance present during the decomposition of calcium carbonate on the product side.

Question 13: Define half-cell.

Answer 13: A half-cell is made up of an electrode structure and conducting electrolyte that is divided by a Helmholtz double layer.

Question 14: What is the role of a salt bridge in an

electrochemical cell?

Answer 14: The role of a salt bridge in an electrochemical cell is that it provides electrical neutrality and prevents the mixing of the electrolytes.

Question 15: Define a redox couple.

Answer 15: A substance that participates in an oxidation and reduction half-reaction is said to be in a redox couple when its reduced and oxidised forms are present together.

Question 16: Explain why

$3Fe_3O_4(s) + 8Al(s) \rightarrow 9Fe(s) + 4Al_2O_3(g)$

is an oxidation reaction?

Answer 16: It is an oxidation reaction because aluminium is getting oxidised. It forms Al2O3 in the product indicating that the addition of oxygen has taken place.

Question 17: The reaction

$Cl_2(g) + 2OH^-(aq) \rightarrow ClO^-(aq) + Cl^-(aq) + H_2O(l)$

represents the process of bleaching. Identify and name the species that bleaches the substances due to their oxidising action.

Answer 17:

$Cl_2(g) + 2OH^-(aq) \rightarrow ClO^-(aq) + Cl^-(aq) + H_2O(l)$

Here, Cl2 is converted into ClO- and Cl-, respectively, through oxidation and reduction. Because Cl- is not an oxidising agent (O.A.). As a result of the hypochlorite ClO- ion's oxidising effect, Cl2 bleaches many materials.

Question 18: Fluorine reacts with ice and results in the change:

$H_2O(s) + F_2(g) \rightarrow HF(g) + HOF(g)$

Justify that this reaction is a redox reaction:

Answer 18:

The oxidation number of each atom involved in the process should be written above its symbol as follows:

The oxidation number of F goes from 0 in F_2 to +1 in HOF, as seen above. Additionally, the number of oxidations decreases from 0 in F_2 to -1 in HF. F is thus both oxidised and reduced in the given reaction. As a result, the given reaction is a redox reaction.

Question 19: MnO_4^{2-} undergoes a disproportionation reaction in an acidic medium but MnO_4^- does not. Give a reason.

Answer 19:

A disproportionation reaction is a redox reaction in which one component with an intermediate oxidation state results in the formation of two molecules with greater and lower oxidation states.

The oxidation states of manganese in their different compounds range from +2 to +7. Disproportionation is not possible with MnO_4^- because of its maximum oxidation state of +7; however, MnO_4^{2-} has a +6 oxidation state and can be both oxidised and reduced.

Question 20: Write the formula for the following compounds:

(a) Mercury(II) chloride

(b) Nickel(II) sulphate

(c) Tin(IV) oxide

(d) Thallium(I) sulphate

(e) Iron(III) sulphate

(f) Chromium(III) oxide

Answer 20: The chemical formula of the given compounds are as follows:

(a) Mercury (II) chloride: $HgCl_2$

(b) Nickel (II) sulphate: $NiSO_4$

(c) Tin (IV) oxide: SnO_2

(d) Thallium (I) sulphate: Tl_2SO_4

(e) Iron (III) sulphate: $Fe_2(SO_4)_3$

(f) Chromium (III) oxide: Cr_2O_3

Question 21: PbO and PbO_2 react with HCl according to the following chemical equations:

$$2PbO + 4HCl \rightarrow 2PbCl_2 + 2H_2O$$

$$PbO_2 + 4HCl \rightarrow PbCl_2 + Cl_2 + 2H_2O$$

Why do these compounds differ in their reactivity?

Answer 21: In the first reaction, none of the atoms' O.N. changes. Hence, it cannot be categorised as a redox reaction. It is an acid-base interaction because PbO is a basic oxide that reacts with HCl acid.

In the second reaction, which is a redox reaction, PbO_2 is reduced and acts as an oxidising agent.

Question 22: Nitric acid is an oxidising agent and reacts with PbO, but it does not react with PbO_2. Explain why.

Answer 22:

PbO, a basic oxide, reacts with HNO_3 in a conventional acid-base manner. On the other hand, lead cannot be further oxidised in PbO_2 because it is in the +4 oxidation state. As a result, no reaction takes place. Consequently, PbO2 is inactive, and only PbO interacts with HNO3.

2PbO + 4HNO3 → 2Pb(NO3)2 + 2H2O

Question 23: The compound AgF2 is an unstable compound. However, if formed, the compound acts as a very strong oxidising agent. Why?

Answer 23: Ag in AgF2 has an oxidation state of +2, which is an unstable oxidation state of Ag. Silver hence readily accepts an electron to generate Ag+ whenever AgF2 is formed. This brings the oxidation state of Ag from +2 to +1, which is more stable. Hence, AgF2 acts as a very strong oxidising agent.

Question 24: Calculate the oxidation number of phosphorus in the following species.

(a) HPO32- and

(b) PO43-

Answer 24:

(a). Let the oxidation number of phosphorus in HPO32- be x.

H + P + 3O2-

⇒ +1 + x + (-2)×3 = -2

⇒ +1 + x − 6 = -2

⇒ x − 5 = -2

⇒ x = − 2 + 5

⇒ x = +3

Thus, the oxidation number of phosphorus in HPO32- is +3.

(b). Let the oxidation number of phosphorus in PO43- be x.

PO_4^{3-}

$\Rightarrow x + 4 \times(-2) = -3$

$\Rightarrow x = -3 + 8 = +5$

$\Rightarrow x = +5$

Thus, the oxidation number of phosphorus in PO_4^{3-} is +5.

Question 25: What sorts of information can you draw from the following reaction?

$(CN)_2(g) + 2OH^-(aq) \rightarrow CN^-(aq) + CNO^-(aq) + H_2O(l)$

Answer 25:

The carbon in $(CN)_2$, CN^- and CNO^- have oxidation numbers +3, +2 and +4, respectively. These are obtained as follows:

Let the oxidation number of C be x.

$(CN)_2$

$2(x - 3) = 0$

$\Rightarrow x = 3$

CN^- $x - 3 = -1$

$\Rightarrow x = 2$

CNO^- $x - 3 - 2 = -1$

$\Rightarrow x = 4$

The oxidation number of carbon in the various species is:

It is clear from the equation that the same compound is being reduced and oxidised at the same time. Disproportionation reactions are those in which the same compound is reduced and oxidised. As a result, it is possible to say that the alkaline decomposition of cyanogen is an illustration of a disproportionation reaction.

Question 26: Identify the redox reactions out of the following reactions and identify the oxidising and reducing agents in them.

(i) $3HCl(aq) + HNO_3(aq) \rightarrow Cl_2(g) + NOCl(g) + 2H_2O(l)$

(ii) $HgCl_2(aq) + 2KI(aq) \rightarrow HgI_2(s) + 2KCl(aq)$

(iii) $Fe_2O_3(s) + 3CO(g) \rightarrow 2Fe(s) + 3CO_2(g)$

(iv) $PCl_3(l) + 3H_2O(l) \rightarrow 3HCl(aq) + H_3PO_4(aq)$

(v) $4NH_3 + 3O_2(g) \rightarrow 2N_2(g) + 6H_2O(g)$

Answer 26:

(i) Putting the oxidation number of each atom, we get:

$3HCl + HNO_3 \rightarrow Cl_2 + NOCl + 2H_2O$

The oxidation number of Cl goes from -1 (in HCl) to 0 (in Cl2). HCl is used as a reducing agent because of the oxidation of Cl–.

HNO3 functions as an oxidising agent because the oxidation number of N reduces from +5 (in HNO3) to +3 (in NOCl).

Thus, reaction (i) is a redox reaction.

(ii) $HgCl_2 + 2Kl \rightarrow Hgl + 2KCl$

There is no change in the oxidation number. Hence this reaction doesn't qualify as a redox reaction.

(iii) $Fe_2O_3 + 3CO \rightarrow 2Fe + 3CO_2$

Fe2O3 works as an oxidising agent because the oxidation number of Fe changes from +3 (in Fe2O3) to 0 (in Fe).

CO works as a reducing agent as the oxidation number of C increases from +2 (in CO) to +4 (in CO2). As a result, it is a redox reaction.

(iv) $PCl_3 + 3H_2O \rightarrow 3HCl + H_2PO_3$

There is no change in the oxidation number of any of the atoms. Hence it is not a redox reaction.

(v) $4NH_3 + 3O_2 \rightarrow 2N_2 + 6H_2O$

Because the oxidation number of N in N_2 grows from -3 to 0, NH_3 functions as a reducing agent.

Furthermore, O_2 functions as an oxidising agent because the oxidation number of O drops from 0 in O_2 to -2 in H_2O.

Hence, this is a redox reaction.

Question 27: Refer to the periodic table given in your book and now answer the following questions:

(a) Select the possible non-metals that can show a disproportionation reaction.

(b) Select three metals that can show a disproportionation reaction.

Answer 27:

One of the reacting compounds in disproportionation reactions always contains an element that can exist in at least three oxidation states,

(a) As the non-metals P, Cl, and S can exist in three or more oxidation states, they can exhibit disproportionation reactions.

(b) Mn, Cu, and Ga can also exhibit such reactions as these metals can exist in three or more oxidation states.

Question 28: Explain redox reactions on the basis of electron transfer. Give suitable examples.

Answer 28:

A chemical reaction in which electrons are transferred between two substances is known as an oxidation-reduction (redox) reaction. An oxidation-reduction reaction is any chemical process in which a molecule, atom, or ion's oxidation number changes as a result of gaining or losing an electron.

As we know, the reactions,

$2Na(s) + Cl_2(g) \rightarrow 2NaCl(s)$

$4Na(s) + O_2(g) \rightarrow 2Na_2(s)$

are redox reactions since they all entail the addition of oxygen or a more electronegative element to sodium.

While sodium, an electropositive element, has been added, chlorine and oxygen are simultaneously being depleted. We also realise that sodium chloride and sodium oxide are ionic compounds, and may be better stated as $Na^+Cl^-(s)$ and Na_2+O_2, respectively, based on our knowledge of chemical bonding (s).

The reactions can be represented as follows:

redox reactions on the basis of electron transfer

One phase of each of the processes above involves electron loss, and the other phase involves electron gain. We can elaborate on one of them, like the production of sodium chloride.

$2Na(s) \rightarrow 2Na^+(g) + 2e^-$

$Cl_2 + 2e^- \rightarrow 2Cl^-(g)$

Since the involvement of electrons is evident, each preceding step is referred to as a half-reaction. The sum of the half-reactions determines the total reaction:

$2Na(s) + Cl_2(g) \rightarrow 2Na + Cl^-(s)$ or $2NaCl(s)$

According to the reactions above, oxidation reactions account for 50% of all reactions involving electron loss. Similar to this, electron gain-related half-reactions are referred to as reduction reactions.

Redox reactions are the fundamental processes of life, including photosynthesis, respiration, combustion, and corrosion or rusting.

Question 29: Arrange the given metals in the order in which they displace each other from the solution of their salts.

(i) Al

(ii) Fe

(iii) Cu

(iv) Zn

(v) Mg

Answer 29:

In a salt solution, a metal with a higher reducing power pushes out metal with lower reducing power.

The following list of metals is in increasing order of their reducing power:

Cu < Fe < Zn < Al < Mg

Thus, Mg can displace Al from its salt solution, but Al cannot displace Mg. Therefore, we can conclude that the order in which the given metals displace each other from the solution of their salts is as given below: Mg >Al>Zn> Fe >Cu

Question 30: Why does fluorine not show a disproportionation reaction?

Answer 30:

A disproportionation reaction is a redox reaction in which one component with an intermediate oxidation state results in the formation of two molecules with greater and lower oxidation states.

$2F_2(g) + 2OH^-(aq) \rightarrow 2F^-(aq) + OF_2(g) + H_2O(l)$

Such a redox reaction cannot take place unless the element is in at least three oxidation states. As a result, during the disproportionation reaction, that element is in the intermediate state and is capable of switching between higher and lower oxidation levels.

Of all the halogens, fluorine is the most electronegative and oxidising element. It is also the smallest.

The disproportionation reaction does not occur, and it only has one positive oxidation state.

Question 31: Which method can be used to find out the strength of the reductant/oxidant in a solution? Explain with an example.

Answer 31:

The relative electrode potential can be determined when a reductant (reducing agent) or oxidant (oxidising agent) is connected in a solution using a cell. The element under discussion can be used as an electrode in a typical cell with a known electrode potential. If the electrode of the given species is positive, it acts as a reductant; if it is negative, it acts as an oxidant.

Let us take the example of Fe^{3+}/Fe with a Standard Hydrogen electrode (SHE). For Fe and H, the half-cell reaction is as follows:

$H^+ + e^- \rightarrow H_2 \quad E^o = 0.0V$

$Fe^{3+} + e^- \rightarrow Fe^{2+} \quad E^o = 0.77$

Any element that requires evaluation can be employed in SHE as an electrode. The amount of emf an element produces in a cell is referred to as its potential.

Eocell = Eocathode – Eoanode

Eocell = 0 – Eoanode

Eocell = 0 – 0.77

Eocell = -0.77

The Fe3+ ion is more likely to go through reduction than hydrogen is. The strength of Fe as a reductant can then be determined by altering the previously believed Fe anode arrangement. As a result, the strength of the oxidant can be identified.

COMPETITIVE CORNER

NEET

JEE

1. Given : $XNa_2HAsO_3 + YNaBrO_3 + ZHCl \rightarrow NaBr + H_3AsO_4 + NaCl$

The values of X, Y and Z in the above redox reaction are respectively :

(1) 2, 1, 3

(2) 3, 1, 6

(3) 2, 1, 2

(4) 3, 1, 4

Solution:

The balanced equation is given below.

$3Na_2HAsO_3 + NaBrO_3 + 6HCl \rightarrow NaBr + 3H_3AsO_4 + 6NaCl$

The value of X, Y and Z are 3, 1 and 6 respectively.

Hence option (2) is the answer.

2. An alkali is titrated against an acid with methyl orange as an indicator, which of the following is a correct combination?

	Base	Acid	Endpoint
1	strong	strong	Pinkish red to yellow
2	weak	strong	Yellow to pinkish-red
3	strong	strong	Pink to colourless
4	weak	strong	Colourless to pink

Solution:

When methyl orange is added to a weak base solution, the solution becomes yellow. When the solution is titrated with a strong acid, after the endpoint, the solution is acidic. So the solution becomes pinkish red.

Hence option (2) is the answer.

3. Consider the following reaction:

$xMnO_4^- + yC_2O_4^{2-} + zH^+ \rightarrow xMn^{2+} + 2yCO_2 + (z/2)H_2O$

The values of x, y and z in the reaction are respectively :-

(1) 5, 2 and 16

(2) 2, 5 and 8

(3) 2, 5 and 16

(4) 5, 2 and 8

Solution:

The balanced equation is given below.

$2MnO_4^- + 5C_2O_4^{2-} + 16 H^+ \rightarrow 2Mn^{2+} + 10CO_2 + 8H_2O$

The values of x, y and z are 2, 5 and 16, respectively.

Hence option (3) is the answer.

4. Consider the reaction

$H_2SO_3(aq) + Sn^{4+}(aq) + H_2O(l) \rightarrow Sn^{2+}(aq) + HSO_4^-(aq) + 3H^+(aq)$

Which of the following statements is correct?

(1) H_2SO_3 is the reducing agent because it undergoes oxidation

(2) H_2SO_3 is the reducing agent because it undergoes reduction

(3) Sn^{4+} is the reducing agent because it undergoes oxidation

(4) Sn^{4+} is the oxidizing agent because it undergoes oxidation

Solution:

Oxidation is the loss of electrons during a reaction by a molecule. In the given equation, H_2SO_3 is the reducing agent because it undergoes oxidation.

Hence option (1) is the answer.

5. In which of the following reaction H_2O_2 acts as a reducing agent ?

(1) $H_2O_2 + 2H^+ + 2e^- \rightarrow 2H_2O$

(2) $H_2O_2 - 2e^- \rightarrow O_2 + 2H^+$

(3) $H_2O_2 + 2e^- \rightarrow 2OH^-$

(4) $H_2O_2 + 2OH^- - 2e^- \rightarrow O_2 + 2H_2O$

(1) (1), (3)

(2) (2), (4)

(3) (1), (2)

(4) (3), (4)

Solution:

Reducing agent is an element or compound that loses an electron to an electron recipient in a redox chemical reaction. In (2) and (4), H_2O_2 acts as a reducing agent.

Hence option (2) is the answer.

6. How many electrons are involved in the following redox reaction?

$Cr_2O_7^{2-} + Fe^{2+} + C_2O_4^{2-} \rightarrow Cr^{3+} + Fe^{3+} + CO_2$ (Unbalanced)

(1) 3

(2) 4

(3) 5

(4) 6

Solution:

A redox reaction is any chemical reaction in which the oxidation number of a molecule, atom, or ion changes by gaining or losing an electron. Chromium and iron are involved in the reaction which is oxidised and reduced. So, a total of 6 electrons are involved in this redox reaction.

Hence option (4) is the answer.

7. Which of the following reactions is an example of a redox reaction ?

(1) $XeF_4 + O_2F_2 \rightarrow XeF_6 + O_2$

(2) $XeF_2 + PF_5 \rightarrow [XeF]^+ PF_6^-$

(3) $XeF_6 + H_2O \rightarrow XeOF_4 + 2HF$

(4) $XeF_6 + 2H_2O \rightarrow XeO_2F_2 + 4HF$

Solution:

In equation (1) Xe undergoes oxidation and oxygen undergoes reduction.

Hence option (1) is the answer.

8. Which of the following is a redox reaction ?

(1) NaCl + KNO3 → NaNO3 + KCl

(2) CaC2O4 + 2HCl → CaCl2 + H2C2O,

(3) Mg(OH)2 + 2NH4Cl → MgCl2 + 2NH4OH

(4) Zn + 2AgCN → 2Ag + Zn(CN)2

Solution:

A redox reaction is any chemical reaction in which the oxidation number of a molecule, atom, or ion changes by gaining or losing an electron. The oxidation state shows a change only in a reaction between zinc and cyanide.

Hence option (4) is the answer.

9. When $KMnO_4$ acts as an oxidising agent and ultimately forms [MnO_4^{2-}, MnO_2, Mn_2O_3 and Mn^{+2}. Then the number of electrons transferred in each case respectively is

(1) 4, 3, 1, 5

(2) 1, 5, 3, 7

(3) 1, 3, 4, 5

(4) 3, 5, 7, 1

Solution:

The oxidation number of Mn in $KMnO_4$, MnO_4^{2-}, MnO_2, Mn_2O_3 and Mn^{+2} 7, 6, 4, 3 and 2 respectively. The number of electrons transferred corresponds to the change in the oxidation number. When $KMnO_4$ acts as an oxidising agent and ultimately forms MnO_4^{2-}, MnO_2, Mn_2O_3 and Mn^{+2}, then the number of electrons transferred in each case are 1,3,4,5 respectively.

Hence option (3) is the answer.

10. For the redox reaction: $Zn_{(s)} + Cu^{2+}$ **(0.1 M)** $\rightarrow Zn^+$ **(1M) +** $Cu_{(s)}$ **taking place in a cell, E°cell is 1.10 volt. E_{cell} for the cell will be (2.303 RT / F = 0.0591)**

(1) 2.14 V

(2) 1.80 V

(3) 1.07 V

(4) 0.82 V

Solution:

$E_{cell} = E°_{cell} - (0.0591/n) \log(1/0.1)$

$E°_{cell} = 1.10$ V

n = 2

$E_{cell} = 1.10 - (0.0591/2) \log(10)$

$= 1.10 - 0.0295$

$= 1.0705$ V

Hence option (3) is the answer.

11. What would happen when a solution of potassium chromate is treated with an excess of dilute nitric acid?

(1) $Cr_2O_7^{2-}$ and H_2O are formed

(2) $Cr_2O_7^{2-}$ is reduced to +3 state of Cr

(3) $Cr_2O_7^{2-}$ is oxidises to +7 state of Cr

(4) Cr^{3+} and $Cr_2O_7^{2-}$ are formed

Solution:

Dilute HNO_3 is an oxidising agent.

$2K_2CrO_4 + 2HNO_3(dil) \rightarrow K_2Cr_2O_7 2KNO_3 + H_2O$

$CrO_4^{2-} + 2HNO_3 (dil) \rightarrow Cr_2O_7^{2-} + 2NO_3^- + H_2O$

Hence option (1) is the answer.

12. Excess of KI reacts with $CuSO_4$ solution and then $Na_2S_2O_3$ solution is added to it. Which of the statements is incorrect for this reaction?

(1) Cu_2I_2 is reduced

(2) Evolved I_2 is reduced

(3) $Na_2S_2O_3$ is oxidized

(4) CuI_2 is formed

Solution:

$2CuSo_2 + 4KI \rightarrow Cu_2I_2 + 2K_2SO_4 + I_2$

$I_2 + 2Na_2S_2O_3 \rightarrow Na_2S_4O_6 + 2NaI$

Here statement (4) is incorrect.

Hence option (4) is the answer.

13. The highest electrical conductivity of the following aqueous solutions is of

(1) 1 M acetic acid

(2) 1 M chloroacetic acid

(3) 1 M fluoroacetic acid

(4) 1 M difluoroacetic acid

Solution:

More the acidity more will be the tendency to release protons. So lighter will be the electrical conductivity. Difluoroacetic acid will be the strongest acid because of the electron-withdrawing effect of two fluorine atoms so as it will show maximum electrical conductivity.

Hence option (4) is the answer.

14. Amount of oxalic acid present in a solution can be determined by its titration with $KMnO_4$ solution in the presence of H_2SO_4. The titration gives unsatisfactory result when carried out in the presence of HCl because HCl

(1) gets oxidised by oxalic acid to chlorine

(2) furnishes H^+ ions in addition to those from oxalic acid

(3) reduces permanganate to Mn^{2+}

(4) Oxidises oxalic acid to carbon dioxide and water

Solution:

HCl is a strong reducing agent. It reduces permanganate to Mn^{2+}.

Hence option (3) is the answer.

15. The oxidation state of chromium in the final product formed by the reaction between KI and acidified potassium dichromate solution is

(1) +4

(2) +6

(3) +2

(4) +3

Solution:

$K_2Cr_2O_7 + 7H_2SO_4 + 6KI \rightarrow Cr_2(SO_4) + 3I_2 + 7H_2O + 4K_2SO_4$

Cr get reduced from +6 Oxidation state to +3 oxidation state.

Hence option (4) is the answer.

HYDROGEN

Question 1. What are metallic/interstitial hydrides? How do they differ from molecular hydrides?

Answer 1:

A hydride is a binary chemical made up of an element and an atom of Hydrogen.

Metallic hydrides are also known as interstitial hydrides. Compounds called metallic/interstitial hydrides are formed when transition metals and Hydrogen are bound together. They are the source of many d-Block and f-Block components. These hydrides are conductors of both heat and electricity.

Molecular hydrides are hydrides that contain additional electronegative atoms bonded to the H-atom. Molecular hydrides have a low electrical conductivity compared to metallic hydrides. While molecular hydrides exist in a gaseous state, metallic hydrides only exist in a solid state.

Interstitial or metallic hydrides are created when hydrogen bonds with transition metals; molecular hydrides, on the other hand, also contain hydrogen bonds with an electronegative atom.

Question 2: Why does Hydrogen occur in a diatomic form rather than a monoatomic form under normal conditions?

Answer: 2 The hydrogen atom has a higher ionisation enthalpy. As a result, taking its electron out is more difficult. It thus tends to exist in low monoatomic form. Instead,

Hydrogen forms a covalent bond with another hydrogen atom to form a diatomic molecule.

Question 3: If the same mass of liquid water and a piece of ice is taken, then why is the density of ice less than that of liquid water?

Answer 3:

Mass per unit volume, or mass/volume, is the unit used to measure density. Water expands as it freezes; therefore, there is more ice than liquid water for a given amount of water. In other words, ice floats on water because it has a lower density than liquid water.

Question 4: What do you understand by the term "non-stoichiometric hydrides"? Do you expect this type of hydride to be formed by alkali metals? Justify your answer.

Answer 4: Hydrogen-deficient compounds are known as non-Stoichiometric hydrides. These are created when dihydrogen reacts with d-block and f-block elements. They do not adhere to the law of constant composition in these hydrides.

E.g.: $LaH_{2.87}$, $YbH_{2.55}$, $TiH\ 1.5 - 1.8$

Alkali metals form stoichiometric hydrides that are naturally ionic. Alkali metal ions and hydride ions are of similar size (208 pm). The hydride ion and the constituent metal experience a significant binding force, forming Stoichiometric hydrides.

Question 5: Complete the following equations:

(i) $PbS(s) + H_2O_2(aq) \rightarrow$

(ii) $CO(g) + 2H_2(g) \xrightarrow{CobaltCatalyst} ?$

Answer 5:

(i) $PbS + 4H_2O_2 \rightarrow PbSO_4 + 4H_2O$ [Redox reaction]

(ii) CO + 2H2 → CH3OH [Redox reaction]

Question 6: How do you expect the metallic hydrides to be useful for hydrogen storage? Explain

Answer 6: Metallic hydrides are deficient in Hydrogen. The law of constant composition does not apply to them.

It has been determined that Hydrogen occupies the interstitial position in the lattices of the hydrides of Pd, Ac, Ni, and Ce, allowing for further hydrogen absorption on these metals.

A significant amount of Hydrogen can be stored in metals like Pt and Pd. Thus, metallic hydrides are employed to store Hydrogen and as an energy source.

Question 7: Give reasons:

(i) Lakes freeze from the top towards the bottom.

(ii) Ice floats on water.

Answer 7:

(i) Due to low temperatures in winter, the lake freezes from top to bottom. Because the cold water is heavier than the warm water, it sinks to the bottom. As it reaches the surface, warm water replaces it. The cycle repeats until the mercury falls below 4 degrees when the lake freezes completely.

(ii) Due to the structure of the ice, it leaves empty spaces between water molecules (i.e. four hydrogen atoms surround one oxygen atom). Thus, ice can float on the water since its density is lower than water.

Question 8: How does the atomic Hydrogen or oxy-hydrogen torch function for cutting and welding purposes? Explain.

Answer 8: Another name for the atomic hydrogen torch is the oxy-hydrogen torch. With the aid of an electric arc, which releases a significant amount of energy, these atoms are created through the dissociation of dihydrogen. The total energy released is 435.88 kJ mol-1. This energy is used to create a temperature of 4000 K, which is necessary for metal cutting and welding. In order to generate a specific temperature on the particular surface that needs to be welded, atomic hydrogen torches are utilised for this purpose.

Question 9: What do you understand by the term 'auto protolysis of water? What is its significance?

Answer 9:

Two water molecules are converted into hydronium ions and hydroxide ions are known as the autoprotolysis of water. This type of self-ionisation occurs in water.

$2H_2O \rightarrow H_3O^+ + OH^-$

The reaction shows the amphoteric characteristic of water. It can act as a base as well as an acid. Electrons are supplied by one water molecule and absorbed by the other.

Question 10: Describe the structure of the common form of ice.

Answer 10: Ice is typically water in its crystalline state. If it crystallises at atmospheric pressure, it takes on a visible hexagonal shape. It condenses to a cubic shape at very low temperatures.

The Structure of Ice

In the 3 – D ice structure, hydrogen bonds and a highly organised structure are present. At a distance of 276 pm, four oxygen atoms form a tetrahedron around each individual oxygen atom. Ice's structure also includes large gaps that can accommodate specific-sized molecules.

Question 11: Discuss briefly the de-mineralisation of water by ion exchange resin.

Answer 11:

De-mineralisation of water is the process of removing all soluble salts from it through cation and anion exchange. In the cation exchange process, sodium, calcium, and magnesium cations change places with hydrogen cations. During the anion exchange procedure, OH is exchanged. These two elements work together to produce water.

$$H^+ + OH^- \rightarrow H_2O$$

Question 12: What causes the temporary and permanent hardness of water?

Answer 12: Hardness in water persists because it contains soluble magnesium and calcium salts in the form of chlorides.

Hardness in water is only temporary since it contains soluble calcium and magnesium salts in the form of hydrogen carbonates.

Question 13: Molecular hydrides are classified as electron deficient, electron precise and electron-rich compounds. Explain each type with two examples.

Answer 13:

Electron deficient: A compound is said to be electron-deficient if it lacks enough electrons for the central atom's octet to be completed. These compounds lack the necessary electrons to form the common electron-pair bonds between each pair of connected atoms.

Examples: Compounds like B_2F_6 and Al_2Cl_6 have less than 8 electrons in their valence shells.

Electron precise: Electron-precise hydrogen compounds have enough valence electrons to form covalent bonds. Those hydrides that possess the precise number of electrons required to form a covalent bond are said to be electron-precise hydrides. These compounds are frequently made with group 14 components. The compounds often take the shape of tetrahedra.

Examples: CH4 and SiH4

Electron rich: Hydrides that contain more electrons than are necessary for bonding are said to be electron-rich hydrides. Most of the additional electrons come from the lone pair of electrons on the core atom. The majority of the components in these compounds come from groups 15, 16, and 17.

Examples: NH3 and PH3

Question 14: Is demineralised or distilled water useful for drinking purposes? If not, how can it be made useful?

Answer 14: We cannot survive without water. It contains a large number of dissolved nutrients that are essential to both people and plants and animals. Demineralised water cannot be used for drinking since it is devoid of any soluble minerals.

This water can be used when the desired minerals are added in the precise amounts needed for growth.

Question 15: Write one chemical reaction for the preparation of D2O2.

Answer 15:

D2O2 can be prepared when D2SO4 is dissolved in water and then reacts with BaO2.

The chemical reaction is as follows: BaO2 + D2SO4 → BaSO4 + D2O

Question 16: Describe the usefulness of water in the biosphere and biological systems.

Answer 16: The usefulness of water in the biosphere and biological systems are as follows:

Water is essential for all life forms, including plants and humans, which together make up about 65% of the human body. It is essential to the biosphere because of its Thermal conductivity, Dipole moment, Specific heat, Dielectric constant and surface tension.

For regulating the atmospheric climate and the body temperatures of all living things, the heat capacity and the heat of vapourisation are quite helpful.

It serves as a carrier of many nutrients that both plants and animals need for a variety of metabolic processes.

Question 17: (i) Draw the gas phase and solid phase structure of H_2O_2.

(ii) H_2O_2 is a better oxidising agent than water. Explain.

Answer 17:

(i) The structure of gas and solid phases of H2O2 have slight variations.

Gas Phase and Solid Phase

(ii) H2O2 is a more effective oxidising agent than water because it changes an acidified KI solution into I2, which gives the starch solution a blue colour but not water. In addition, H2O2 is the only substance that can turn black PbS into white PbSO4, not water.

Question 18: Knowing the properties of H2O and D2O, do you think that D2O can be used for drinking purposes?

Answer 18: D2O is referred to as "heavy water" and serves as a moderator (slows down the rate of reaction). This characteristic prevents it from being used for drinking because it slows down catabolic and anabolic reactions. These changes in the body may cause a casualty.

Question 19: Dihydrogen reacts with dioxygen (O_2) to form water. Write the name and formula of the product when the isotope of Hydrogen, which has one proton and one neutron in its nucleus, is treated with oxygen. Will the reactivity of both the isotopes be the same towards oxygen? Justify your answer.

Answer 19:

One proton and one neutron make up the hydrogen isotope deuterium (D). Dideuterium interacts with dioxygen to produce deuterium oxide, sometimes known as heavy water.

With oxygen, H2 and D2 will react in very different ways. H2 is more reactive than D2 in oxygen reactions because the D-D bond is stronger than the H-H bond.

Question 20: What is the difference between the terms 'hydrolysis' and 'hydration'?

Answer 20: The difference between the terms 'hydrolysis' and 'hydration' is as follows:

Hydration: It is the process by which one or more molecules are added to a molecule or an ion to produce hydrated compounds.

CuSO4 + 5H2O → CuSO4.5H2O

Hydrolysis: It is the name for a chemical reaction in which a substance reacts with hydroxide ions and Hydrogen from water molecules to produce products.

NaH + H2O → NaOH + H2

Question 21: Explain why HCl is a gas and HF is a liquid.

Answer 21:

Intermolecular hydrogen bonds join HF molecules. As a result, HF is a liquid at room temperature. Since the HCl molecules lack intermolecular hydrogen bonds, it is a gas at room temperature.

Question 22: How can saline hydrides remove traces of water

from organic compounds?

Answer 22: Saline hydrides are ionic by nature. Water and saline hydrides react, releasing hydrogen gas and forming metal hydroxide in the process. When introduced to an organic solvent, they cause the water to react.

$$AH(S) + H_2O(l) \rightarrow AOH(aq) + H_2(g)$$

The metallic hydroxide is left behind as the Hydrogen escapes into the environment. The dry organic solvent separates over.

Question 23: When the first element of the periodic table is treated with dioxygen, it gives a compound whose solid-state floats on its liquid state. This compound has the ability to act as an acid as well as a base. What products will be formed when this compound undergoes autoionisation?

Answer 23:

The first element in the periodic table is H, and its molecule is dihydrogen (H_2). Dihydrogen reacts with dioxygen to form water. Water is fluid when it's at a normal temperature. As water freezes, it expands to form ice. In other words, ice floats on top of liquid water because it is less dense than water.

In its natural state, water is amphoteric, acting as a base in the presence of strong acids and acids when strong bases are present.

auto-protolysis.

It happens that conjugate bases and acids are formed. This process of water self-ionisation is known as auto-protolysis.

Question 24: How does H_2O_2 behave as a bleaching agent?

Answer 24: In both basic and acidic conditions, hydrogen peroxide functions as an effective oxidising agent. When introduced to the fabric, it breaks down the chromophores' (colour-producing agents) chemical bonds. As a result, the cloth becomes whiter, and the visible light is not absorbed.

Question 25: Hydrogen generally forms covalent compounds. Give a reason.

Answer 25:

Hydrogen tends to form covalent compounds because it shares its electron with other elements. Hydrogen is an element with an atomic number of one and only one electron. Since it cannot lose an electron, it prefers to share electrons with other atoms, forming covalent compounds.

Question 26: Why is water an excellent solvent for ionic or polar substances?

Answer 26: Water functions as a polar solvent due to its high dielectric constant. The high dielectric constant of water reduces the attraction between cation and anion. Therefore, water molecules can easily remove ions from the lattice site using dipole forces.

Question 27: Why is the Ionisation enthalpy of Hydrogen higher than that of sodium?

Answer 27:

The H atom has a higher nuclear attraction than the Na atom; hence it takes more energy to remove a valence electron from it. Because of this, Hydrogen's ionisation enthalpy (1312 kJ/mol) is higher than sodium's (496 kL/mol).

Question 28: Which fuel is used as rocket fuel?

Answer 28: Since Hydrogen is both light and incredibly strong, it is employed as rocket fuel. It burns extremely hot and has the smallest molecular weight of any known chemical.

Question 29: What is the importance of heavy water?

Answer 29:

In nuclear reactors, heavy water is used as a neutron moderator to slow down neutrons and increase the likelihood that they will interact with fissile uranium-235 rather than the neutron-collecting uranium-238.

In studying reaction mechanisms and manufacturing other deuterium compounds like CD4, D2SO4, and others, it is used as a tracer chemical.

Question 30: What is the behavioural similarity between NH3, H2O, and HF compounds?

Answer 30: They act as electron donors or Lewis bases. Lone pairs on highly electronegative atoms like N, O, or F in hydrides cause hydrogen bonds between molecules.

Question 31: With the help of suitable examples, explain the property of H_2O_2 that is responsible for its bleaching action.

Answer 31:

The unstable oxygen that H2O2 produces after its breakdown causes it to bleach materials.

$$H2O2 \rightarrow H2O + [O]$$

A coloured substance that has been exposed to nascent oxygen becomes colourless. Feathers, silk, wool, ivory, and other materials can all be bleached using this method. It can be used to bleach paper, oils, and fats.

Question 32: What is the pH of water?

Answer 32: pH determines the concentration of hydrogen ions (H+) in a solution. In pure water, random processes that produce the ions H+ and OH– tend to produce ions. The amount of H+ generated in pure water is roughly similar to an OH–. Because of this, a pH of 7 is regarded as neutral.

Question 33: Why is hydrogen peroxide stored in wax-lined bottles?

Answer 33:

The rough surfaces of glass break down hydrogen peroxide, any alkali oxides present and light. H_2O_2 is often stored in coloured Teflon or plastic bottles covered with paraffin wax to avoid decomposition.

Question 34: Why is dihydrogen gas not preferred in balloons?

Answer 34: Dihydrogen should have been used in balloons because it is the lightest gas. However, it is not advised due to its combustibility.

Question 35: How will you account for the 104.5° bond angle in water?

Answer 35:

Two lone pairs of electrons exist after two of an oxygen atom's six electrons are linked to a hydrogen atom. These lone pairs of electrons provide the bond angle in H_2O with a value of 104.5°. This can be explained by the valence shell electron pair repulsion concept (VSEPR).

The sp3 hybridisation of the oxygen in the H_2O molecule produces the tetrahedral structure. Lone pairs occupy two spots, and H atoms occupy two positions by forming sigma bonds with two hybrid orbitals. The actual bond angle is 104.5°, as opposed to the expected bond angle of 109.5°. In comparison to bond pairs, lone pairs are more repellent to one another. The result is a decrease in the bending angle of water from 109.5° to 104.5°.

Question 36: Atomic hydrogen combines with almost all elements but molecular Hydrogen does not. Explain.

Answer 36:

Atoms of Hydrogen are incredibly brittle. Atomic Hydrogen needs one more electron to complete its electronic configuration and attain stability because it is in the 1s1 state. Atomic Hydrogen is particularly reactive as a result, reacting with almost every element. However, it interacts in three different ways: first, by losing one electron to H+, second, by gaining one electron from H+, and third, by creating single covalent bonds by exchanging electrons with other atoms.

On the other hand, the H-H bond has a very high bond dissociation energy (435.88 kJ/mol1). As a result, only a few elements can react with molecular Hydrogen at room temperature.

COMPETITIVE CORNER

NEET

1. Electrolysis of brine produces

   a. chlorine gas
   b. hydrogen gas
   c. sodium hydroxide
   d. all the above

Answer: (d)

2. A reactant containing the element that is oxidised is called

   a. reducing agent
   b. oxidising agent
   c. hydrogen
   d. sublime

Answer: (a)

3. By losing one or two electrons the atoms of metal are

   a. oxidised
   b. reduced
   c. hydrogenated
   d. sublimated

Answer: (a)

4. Electrolytes conduct electricity in

   a. solid state
   b. liquid state
   c. gaseous state
   d. plasma state

Answer: (b)

5. Loss of hydrogen atoms by an element is called

   a. hydrogenation
   b. oxidation
   c. reduction
   d. sublimation

Answer: (b)

6. The electrolyte among the following is

   a. NaOH
   b. Urea

c. glucose

d. benzene

Answer: (a)

7. O-O-H bond angle in H2O2 is

    a. 97°

    b. 106°

    c. 120°

    d. 109°28'

Answer: (a)

8. Which of the following is very high for proton?

    a. radius

    b. ionization potential

    c. charge

    d. hydration energy

Answer: (d)

9. The list which contains only elements is

    a. air, water, oxygen

    b. hydrogen, oxygen, brass

    c. air, water, fire, earth

    d. calcium, sulphur, carbon

Answer: (d)

10. The smallest part of an element that cannot exist as

a free state is

   a. ion
   b. charge
   c. atom
   d. molecule

Answer: (c)

JEE

1. Identify the incorrect statement regarding heavy water.

(1) It reacts with SO_3 to form deuterated sulphuric acid (D_2SO_4).

(2) It is used as a coolant in nuclear reactors.

(3) It reacts with CaC_2 to produce C_2D_2 and $Ca(OD)_2$.

(4) It reacts with Al_4C_3 to produce CD_4 and $Al(OD)_3$.

Solution:

Heavy water is used as a moderator in nuclear reactors to control the speed of neutrons. It is not used as a coolant.

Hence option (2) is the correct answer.

2. Determine the total number of neutrons in three isotopes of hydrogen

(1) 1

(2) 2

(3) 3

(4) 4

Solution:

Number of neutrons = 0+1+2 = 3

Hence option (3) is the answer.

3. Hydrogen peroxide acts both as an oxidising and as a reducing agent depending upon the nature of the reacting species. In which of the following cases does H_2O_2 act as a reducing agent in acid medium?

(1) MnO_4^-

(2) SO_3^{2-}

(3) KI

(4) $Cr_2O_7^{2-}$

Solution:

Reducing agent is an element or compound that loses or donates an electron to an electron recipient, oxidizing agent in a redox chemical reaction.

$$H_2O_2 + MnO_4^- \rightarrow Mn^{+2} + O_2$$

Hence option (1) is the answer.

4. Hydrogen peroxide oxidises $[Fe(CN)_6]^{4-}$ to $[Fe(CN)_6]^{3-}$ in acidic medium but reduces $[Fe(CN)_6]^{3-}$ to $[Fe(CN)_6]^{4-}$ in alkaline medium. The other products formed are, respectively :

(1) $(H_2O + O_2)$ and $(H_2O + OH^-)$

(2) H_2O and $(H_2O + O_2)$

(3) H_2O and $(H_2O + OH^-)$

(4) $(H_2O + O_2)$ and H_2O

Solution:

$[Fe(CN)_6]^{4-}$ reacts with hydrogen peroxide in acidic medium to form $[Fe(CN)_6]^{3-}$ and water.
$$[Fe(CN)_6]^{4-} + H_2O_2 + 2H^+ \rightarrow [Fe(CN)_6]^{3-} + 2H_2O$$
$[Fe(CN)_6]^{3-}$ reacts with hydrogen peroxide in alkaline medium to form $[Fe(CN)_6]^{4-}$, oxygen and water.
$$[Fe(CN)_6]^{3-} + H_2O_2 + 2OH^- \rightarrow [Fe(CN)_6]^{4-} + O_2 + 2H_2O$$

Hence option (2) is the answer.

5. Which of the following statements are correct?

(1) On decomposition of H_2O_2, O_2 gas is released.

(2) 2-ethylanthraquinol is used in the preparation of H_2O_2

(3) On heating $KClO_3$, $Pb(NO_3)_2$, $NaNO_3$, O_2 gas is released.

(4) In the preparation of sodium peroxoborate, H_2O_2 is treated with sodium metaborate.

(1) 1,2,4

(2) 2,3,4

(3) 1,2,3,4

(4) 1,2,3

Solution:

All the given statements are correct.

Hence option (3) is the answer.

6. The isotopes of hydrogen are

(1) protium, deuterium and tritium

(2) protium and deuterium only

(3) deuterium and tritium only

(4) tritium and protium only.

Solution:

The isotopes of hydrogen are protium, deuterium and tritium.

Hence option (1) is the answer.

7. The synonym for water gas, when used in the production of methanol, is

(1) fuel gas

(2) natural gas

(3) laughing gas

(4) syn gas.

Solution:

Water-gas is $CO+H_2$.

It is used for the synthesis of methanol. So it is called syn gas.

Hence option (4) is the answer.

8. The chemical nature of hydrogen peroxide is

(1) oxidising and reducing agent in both acidic and basic medium

(2) oxidising agent in acidic medium, but not in basic medium

(3) oxidising and reducing agent in acidic medium, but not in basic medium

(4) reducing agent in basic medium, but not in acidic medium.

Solution:

Hydrogen peroxide acts as both oxidising and reducing agent in both acidic and basic medium.

Hence option (1) is the answer.

9. The correct statements among (A) to (D) regarding H_2 as a fuel are

(A) It produces less pollutants than petrol.

(B) A cylinder of compressed dihydrogen weighs ~30 times more than a petrol tank producing the same amount of energy.

(C) Dihydrogen is stored in tanks of metal alloys like NaNi5.

(D) On combustion, values of energy released per gram of liquid dihydrogen and LPG are 50 and 142 kJ, respectively.

(1) (A), (B) and (C) only

(2) (B), (C) and (D) only

(3) (A) and (C) only

(4) (B) and (D) only

Solution:

On combustion, the value of energy released per gram of liquid dihydrogen is 142kJ.

The energy released per gram of LPG is 50kJ. Statement D is wrong.

Hence option (1) is the answer.

10. NaH is an example of

(1) metallic hydride

(2) saline hydride

(3) electron-rich hydride

(4) molecular hydride.

Solution:

NaH is an example of saline hydride. Hydrides are binary compounds of the elements with hydrogen.

Hence option (2) is the answer.

11. Very pure hydrogen (99.9%) can be made by which of the following processes?

(1) Reaction of salt like hydrides with water

(2) Reaction of methane with steam

(3) Mixing natural hydrocarbons of high molecular weight

(4) Electrolysis of water

Solution:

Very pure hydrogen (99.9%) can be made by electrolysis of water using platinum electrodes in presence of small amount of acid or alkali.

$2H_2O_{(l)} \rightarrow 2H_{2(g)} + O_2$

Hence option (4) is the answer.

12. Which physical property of dihydrogen is wrong?

(1) Colourless gas

(2) Odourless gas

(3) Tasteless gas

(4) Non-inflammable gas

Solution:

Dihydrogen is an inflammable gas.

Hence option (4) is the correct answer.

13. The metal that gives hydrogen gas upon treatment with both acid, as well as base, is (1) magnesium

(2) zinc

(3) mercury

(4) iron.

Solution:

Zinc is the metal that gives hydrogen gas upon treatment with both bases as well as an acid.

Hence option (2) is the correct answer.

14. Which one of the following statements about water is false?

(1) Water is oxidized to oxygen during photosynthesis.

(2) Water can act both as an acid and as a base.

(3) There is extensive intramolecular hydrogen bonding in the condensed phase.

(4) Ice formed by heavy water sinks in normal water.

Solution:

There is extensive intermolecular hydrogen bonding in water molecules in the condensed phase. It is not intramolecular hydrogen bonding.

Hence option (3) is the correct answer.

15. Which one of the following processes will produce

hard water?

(1) Saturation of water with $CaCO_3$

(2) Saturation of water with $MgCO_3$

(3) Saturation of water with $CaSO_4$

(4) Addition of Na_2SO4 to water

Solution:

Hard water contains calcium and Magnesium salt in the form of hydrogen carbonate, chloride and sulphate. Permanent hardness is introduced when water passes over rocks containing the sulphates or chlorides of both calcium and magnesium.

Hence option (3) is the correct answer.

THE S- BLOCK ELEMENTS

Question 1. The alkali metals have a low melting point. Which among the following alkali metal is expected to melt when the room temperature rises to 30°C?

(i) Na

(ii) K

(iii) Rb

(iv) Cs

Answer 1. (iv) Cs

Option (iv) is the correct answer.

Question 2. Alkali metals react with the water vigorously to form the hydroxides and dihydrogen.

Which among the following alkali metals reacts with the water least vigorously?

(i) Li

(ii) Na

(iii) K

(iv) Cs

Answer 2. (i) Li

Option (i) is the answer.

Question 3. The reducing power of the metal depends on the different factors. Suggest the factor that makes Li one of the strongest reducing agents in the aqueous solution.

(i) Sublimation enthalpy

(ii) Ionisation enthalpy

(iii) Hydration enthalpy

(iv) Electron-gain enthalpy

Answer 3. (iii) Hydration enthalpy

Option (iii) is the answer.

Question 4. The metal carbonates decompose on heating to give the metal oxide and carbon dioxide. Which among the metal carbonates is the most stable thermally?

(i) $MgCO_3$

(ii) $CaCO_3$

(iii) $SrCO_3$

(iv) $BaCO_3$

Answer 4. (iv) $BaCO_3$

Option (iv) is the answer.

Question 5. Which among the following carbonates given is unstable in air and is kept with the CO_2 atmosphere to avoid the decomposition.

(i) $BeCO_3$

(ii) $MgCO_3$

(iii) $CaCO_3$

(iv) BaCO3

Answer 5. (i) BeCO3

Option (i) is the answer.

Question 6. Metals form the basic hydroxides. Which among the following metal hydroxide is the least basic?

(i) Mg(OH)2

(ii) Ca(OH)2

(iii) Sr(OH)2

(iv) Ba(OH)2

Answer 6. (i) Mg(OH)2

Option (i) is the answer

Question 7. Some among the Group 2 metal halides are covalent and soluble in the organic solvents.

From the following metal halides, the one that is soluble in the ethanol is

(i) BeCl2

(ii) MgCl2

(iii) CaCl2

(iv) SrCl2

Answer 7. (i) BeCl2

Option (i) is the answer.

Question 8. The order for the decreasing ionization enthalpy in the alkali metals is

(i) Na > Li > K > Rb

(ii) Rb < Na < K < Li

(iii) Li > Na > K > Rb

(iv) K < Li < Na < Rb

Answer 8. (iii) Li > Na > K > Rb

Option (iii) is the answer.

Question 9. The solubility for the metal halides depends on their nature, the lattice enthalpy and the hydration enthalpy of the individual ions. From the following fluorides of the alkali metals, the lowest solubility of the LiF in water is because of the

(i) Ionic nature of lithium fluoride

(ii) High lattice enthalpy

(iii) High hydration enthalpy for lithium-ion.

(iv) Low ionization enthalpy of the lithium atom

Answer 9. (ii) High lattice enthalpy

Option (ii) is the answer.

Question 10. The amphoteric hydroxides react with both the alkalies and acids. Which among the

following Group 2, metal hydroxides are soluble in the sodium hydroxide?

(i) $Be(OH)_2$

(ii) $Mg(OH)_2$

(iii) $Ca(OH)_2$

(iv) $Ba(OH)_2$

Answer 10. (i) $Be(OH)_2$

Option (i) is the answer.

Question 11. Match the elements correctly given in Column I with their suitable properties mentioned in Column II.

Column I
(i) Li

(ii) Na

(iii) Ca

(iv) Ba

Column II
(a) Insoluble sulphate

(b) Strongest monoacidic base

(c) Most negative E value of the alkali metals.

(d) Insoluble oxalate

(e) $6s^2$

outer electronic configuration

Answer 11.

(i) is c

(ii) is b

(iii) is d

(iv) is a,e

Question 12. Match the compounds correctly given in Column I with their right uses mentioned in Column II.

Column I
(i) $CaCO_3$

(ii) $Ca(OH)_2$

(iii) CaO

(iv) $CaSO_4$

Column II
(a) Dentistry, ornamental work

(b) Manufacturing of sodium carbonate from caustic soda

(c) Manufacturing of the high-quality paper

(d) Used in the whitewashing

Answer 12.

(i) is c

(ii) is d

(iii) is b

(iv) is a

Question 13. Match the elements correctly given in Column I with the appropriate colours they impart for the flame has given in Column II.

Column I
(i) Cs

(ii) Na

(iii) K

(iv) Ca

(v) Sr

(vi) Ba

Column II
(a) Apple green

(b) Violet

(c) Brick red

(d) Yellow

(e) Crimson red

(f) Blue

Answer 13.

(i) is f

(ii) is d

(iii) is b

(iv) is c

(v) is e

(vi) is a

Question 14. For the following questions, a statement of the Assertion (A) is followed by the statement of the Reason (R). Pick the correct option out of the alternatives given below each question.

(I). Assertion (A): The carbonate of the Lithium decomposes easily on heating to form the lithium oxide and CO2.

Reason (R): Lithium being very small in size polarises large carbonate ions leading to the formation of the more stable Li2O and CO2.

(i) Both of them, A and R, are correct, and R is the correct explanation for A.

(ii) Both of them, A and R, are correct, but R is not the correct explanation for A.

(iii) Both of them A and R are not correct

(iv) A is not right, but R is correct.

Answer (I). Option (i) is correct.

(II). Assertion (A): Beryllium carbonate is kept in the atmosphere of the carbon dioxide.

Reason (R): Beryllium carbonate is unstable as well as decomposes to give beryllium oxide and the carbon dioxide.

(i) Both of them, A and R, are correct, and R is the correct explanation for A.

(ii) Both of them, A and R, are correct, but R is not the correct explanation for A.

(iii) Both of them, A and R, are not correct.

(iv) A is not correct, but R is right.

Answer (II). Option (i) is correct.

Question 15. How do you account for the strong reducing power for the Lithium in the aqueous solution?

Answer 15. Lithium has the highest negative $E^\ominus$ value, that is $-3.04V$. Lithium has the small atomic size and the highest ionisation enthalpy, but it is compensated by the high hydration enthalpy. As a result, the reducing power of the Lithium is highest in the aqueous solution.

Question 16. Complete the following reactions given below:

(i) $O_2^{2-} + H_2O \rightarrow$

(ii) $O_2^- + H_2O \rightarrow$

Answer 16. (i) $O_2^{2-} + 2H_2O \rightarrow H_2O_2 + 2OH^-$

(ii) $2O_2^- + 2H_2O \rightarrow H_2O_2 + O_2 + 2OH^-$

Question 17. Discuss the trend for the following:

(i) Thermal stability of the carbonates of the Group 2 elements.

(ii) The solubility and the nature of the oxides of the Group 2 elements.

Answer 17. (i) The thermal stability of the carbonates increases for the increase in the cationic size. The more stable the oxide of the alkaline earth metal, the less stable the carbonate for the same. So, $BeCO_3$ is highly unstable, and the BeO is stable.

(ii) Alkali metals from the oxides with the oxygen and give the metal oxides. The oxides will be a basic exception to BeO as the BeO is amphoteric. They also react with the water to form the sparingly soluble hydroxides when the size of the cations increases, BeO and MgO have the highest lattice energy as well as they are insoluble in the water.

Question 18. Why are the $BeSO_4$ and $MgSO_4$ readily soluble in the water, whereas the $CaSO_4$, $SrSO_4$ and $BaSO_4$ are insoluble?

Answer 18. $BeSO_4$ and $MgSO_4$ are readily soluble in the water, but $CaSO_4$, $SrSO_4$, and $BaSO_4$ are insoluble. This is due to the greater hydration enthalpies of the Be^{2+} and Mg^{2+} ions. They overcome the lattice enthalpy factor, and thus, their sulphate is soluble in the water.

Question 19. All compounds of the alkali metals are easily soluble in the water, whereas the lithium compounds are more soluble in the organic solvents. Explain.

Answer 19. The alkali metal compounds form the ionic compounds because of their large ionic size and low ionization enthalpy, whereas Lithium forms the compounds of the covalent nature as their small ionic size, the high ionization enthalpy, and the high electronegativity.

Question 20. For the Solvay process, can we get the sodium carbonate directly treated with the solution containing the

(NH4) 2CO3 with the sodium chloride? Explain.

Answer 20. For the Solvay process, the carbon dioxide is passed through the concentrated solution of the sodium chloride saturated with the ammonia, which forms ammonium carbonate, followed by the ammonium hydrogen carbonate. Ammonium hydrogen carbonate crystals separate as they are heated to form the sodium carbonate. The NH3 is recovered from the solution that contains the NH4Cl and then is heated when treated with Ca(OH)2. The reaction of the (NH4)2CO3 with the NaCl gives two products, Na2CO3 and NH4Cl, that are both soluble in the water that does not shift the equilibrium to the right.

Question 21. Write Lewis structure of the O-2 ion. Also, find out the oxidation state for each oxygen atom? What is the average oxidation state of the oxygen for this ion?

Answer 21.

Source: Internet

Oxygen atoms with zero charges have six electrons. Thus the oxidation state is 0. If the oxygen atom has a negative charge, it has the seven electrons. Thus, the oxidation state is -1. The average oxidation state is 0 + (-1)/2 = -1/2.

Question 22. What is the structure for the BeCl2 molecule in the gaseous and the solid state?

Answer 22. The gaseous/vapour state is different from the solid state. The structure of the BeCl2 in the solid-state is the polymeric chain structure. BeCl2 tends to form the chloro-bridged dimer at the temperatures below 1200K and dissociates into the linear monomer at the high temperatures of the order of 1200 K.

Question 23. Beryllium and magnesium do not give colour for the flame, but the other alkaline earth metals do so. Why?

Answer 23. The valence electrons get excited to the higher energy level if the alkaline earth metal is heated.

It radiates energy that belongs to the visible region. If this excited electron comes back to the energy level that is low, the colour is seen. The electrons are firmly bound to the beryllium and magnesium. The energy needed to excite the electrons is very high. If the electron reverts back to the original position, the energy released does not fall in the visible region. So, no colour is observed in the flame.

Question 24. Potassium carbonate could be prepared by the Solvay process. Explain why?

Answer 24. The Solvay process is not applicable for the preparation of the potassium carbonate as the potassium carbonate is soluble in the water, and it doesn't precipitate out as the sodium bicarbonate.

Question 25. Draw the structure for the (i) $BeCl_2$ (vapour) (ii) $BeCl_2$ (solid).

Answer 25. (i) $BeCl_2$ has the linear structure and exists as the monomer in the vapour state.

(ii) For the solid phase, $BeCl_2$ is the polymer.

Question 26. Describe the importance for the following : (i) limestone, (ii) cement, and (iii) plaster of Paris.

Answer 26. Uses of the cement:

Construction of the bridge
Plastering
An essential ingredient in the concrete
Uses of the Plaster of Paris:

Used to make the casts and the moulds
Used to make the surgical bandages
Uses of the limestone:

Preparation of the cement and the lime

As a flux in the iron ore smelting

Question 27. Why are the lithium salts commonly hydrated, whereas the other alkali ions are usually anhydrous?

Answer 27. As Lithium has the smallest size for all the alkali metals, it can be easily polarised to water molecules. Therefore, the smaller the size of the ion, the greater is the ability to polarise the water molecules.

Therefore, the trihydrated Lithium Chloride and the other Lithium salts could be easily polarised. Because of this, the other alkali metal ions could only form the anhydrous salts.

Question 28. What happens if:

(i) Sodium metal is immersed in the water?

(ii) Sodium metal is heated in the free supply of the air?

(iii) Sodium peroxide gets dissolved in the water?

Answer 28. (i) Sodium reacts to form the NaOH and H2 gas if it is dropped in the water. The reaction occurs as given below:

$2Na(s) + 2H_2O(l)$

$2NaOH(aq) + H_2(g)$

(ii) Sodium peroxide is formed if sodium reacts with oxygen during heating it in the presence of air. The reaction proceeds as given below:

$2Na(s) + O_2(s)$

$Na_2O_2(s)$

(iii) NaOH and water are formed as the result of the hydrolysis of the Sodium peroxide if it is dissolved in the water.

$Na_2O_2(s) + 2H_2O(l)$

2NaOH(aq) + H2O2(aq)

Question 29. How will you explain the following observations?

(i) BeO is almost insoluble, whereas the BeSO4 is soluble in the water.

(ii) BaO is soluble, whereas BaSO4 is insoluble in the water.

(iii) LiI is more soluble than KI in the ethanol.

Answer 29.

(i) The sizes of the Be2+ and O2- are small and are highly compatible with each other. As a result, the high amount of the lattice energy is released during the formation. The hydration energy if it is made to dissolve in the water is not enough for overcoming the lattice energy. So, BeO is almost insoluble in the water.

The size of the SO42- is large compared to the Be2+, and there is lesser compatibility and lattice energy that can be easily overcome by the hydration energy. Therefore, BeSO4 is easily soluble in the water.

(ii) The sizes of the Ba2+ and SO42- are large and are highly compatible with each other. As a result, the high amount of the lattice energy is released during their formation. The hydration energy if it is made to dissolve in the water is not enough for overcoming the lattice energy. So, BaSO4 is insoluble in the water.

The size of the O2- is small compared to the Ba2+, and there is lesser compatibility and lattice energy that can be easily overcome by the hydration energy. So, BaO is easily soluble in the water.

(iii) The lithium-ion has a smaller size, and as a result, it has the higher polarising capability. This enables it to polarise the electron cloud around the iodide ion, therefore, resulting in the greater covalent character in the LiI than KI. So, LiI is easily soluble in the ethanol.

Question 30. What are the common physical and chemical characteristics of the alkali metals?

Answer 30. Physical properties of the alkali metals:

(1) The alkali metal is soft. Hence, we can cut them easily. We can cut the sodium metal even when using the knife.

(2) Generally, the alkali metal is lightly coloured as well, as mostly they are seen as silvery white.

(3) Its atomic size is larger. Therefore, their density is low. The density of the alkali metal increases as we move down in the group from the Li to Cs, except for the K, which has lower density than the sodium.

(4) Alkali metal has weak metallic bonding. Thus, they have low boiling and melting points.

(5) The salts present in the alkali metals expose colour to the flames as the heat of the flame excites the electrons, which are located on the outer orbital, to the higher energy level. For the excited state, electrons get reversed back to the ground level, and thus the emission of excess energy in the form of the radiation falls in the visible region.

(6) The metals like K and Cs lose electrons if they get irradiated with the light and also display the photoelectric effect.

Chemical properties of the alkali metals:

(1) Alkali metal reacts with the water and forms the oxides and hydroxides. Thus, the reaction will be more spontaneous when moving down the group.

(2) Alkali metal reacts with water as well as forms the dihydrogen and hydroxides.

General reaction:

2M + 2H2O

2M+ + 2OH– + H2

(3) Dihydrogen reacts with the alkali metals and forms the metal hydrides. The hydrides formed have the higher melting points, and they are solids that are ionic.

2M + H2

2M+ H–

(4) Alkali metals directly react with the halogens and form the ionic halides except Li.

2M + CL2

2MCI   (M = Li, K, Rb, Cs)

It has the ability to easily distort the cloud, having the electrons that are around the –ve halide ion as the lithium-ion is smaller in size. So, the Lithium halide is naturally covalent.

(5) The alkali metals have strong reducing agents. This increases as we go down the group exception lithium. Due to the high hydration energy, it results in the strong reducing agent for all the alkali metals.

(6) For the blue-coloured solution(deep blue) that is naturally conducting, it gets dissolved in the liquid ammonia.

M +(x + y) NH3

[M (NH3)x]+ + [e (NH3)y]–

Question 31. Discuss the general features and trends in the properties of the alkaline earth metals

Answer 31. General features:

(i) (Noble gas) ns2 is the electronic configuration of the alkaline earth metal.

(ii) To attain the nearest inert gas configuration, these metals lose two of the electrons. So its oxidation state is +2.

(iii) The ionic radii and atomic radii are smaller in comparison with the alkali metals. When they moved down in the group, there was an increase in the trend of the ionic radii and atomic radii due as a result of the decrease in the effective nuclear charge.

(iv) The ionisation enthalpy is low as the alkaline earth metals are larger in size. The first ionisation enthalpy is higher in comparison with the metals of group 1.

(v) They appear as lustrous and silvery-white. They are soft like the alkali metals.

(vi) Factors that lead for the alkaline earth metals to have a high boiling point and melting points:

(*) Atoms of alkali metals are larger than in comparison with the alkaline earth metals.

(*) The strong metallic bonds are formed with the two valence electrons.

(vii) Ca- brick red; Sr- crimson red; Ba-apple green gives colours of the flames.

The electrons are bound strongly in the elements Be and Mg. Therefore, they do not give any colours for the flame.

The alkali metals are more reactive than the alkaline earth metals.

Chemical properties:

(i) Reaction with the air and water: Due to the formation of the oxide layer on the surface, beryllium and magnesium are most inert to the water and air.

(a) BeO and Be3 N2 are formed if powdered Be is burnt in the air.

(b) In the formation of MgO and Mg3 N2, Mg is burnt when exposed to the air with dazzling sparkle as Mg is more electropositive.

(c) The formation of their respective nitrides and oxides is by the instant reaction of Sr, Ca, and Ba with air.

(d) Ca, Sr, and Ba could react vigorously even with the water, which is cold.

(ii) When they react with the halogens, halides are formed at high temperature.

M +X2

MX2 (X = F,CL,Br,I)

(iii) Exception Be, all the alkaline earth metals react with the hydrogen to form hydrides.

(iv) The alkaline earth metals instantly react with the acids to form salts along with the liberation of hydrogen gas.

M +2HCl

MCl2 + H2(X)

(e) Reducing Nature: The alkaline earth metals are the strong reducing agents as the alkali metals, but the reducing power is less if compared to the alkali metals. In general, the reducing character increases from the top to the bottom.

(f) Solutions in the liquid ammonia: The alkaline earth metals dissolve in the liquid ammonia to give deep blue-black solutions like the alkali metal.

COMPETITIVE CORNER

NEET

1. Which of the compounds is known as Slaked lime?

(a) CaO

(b) $CaSO_4$

(c) $Ca(OH)_2$

(d) $CaCO_3$

Answer: (c)

2. Which of the ions have maximum hydration energy?

(a) Sr^{2+}

(b) Ca^{2+}

(c) Mg^{2+}

(d) Be^{2+}

Answer: (d)

3. As compared to K, Na has

(a) higher ionization potential

(b) lower melting point

(c) lower electronegativity

(d) larger atomic radius

Answer: (a)

4. Which one is the most stable carbonate?

(a) $BaCO_3$

(b) $MgCO_3$

(c) $CaCO_3$

(d) $BeCO_3$

Answer: (a)

5. Plaster of Paris (POP) is

(a) $CaSO_4\, H_2O$

(b) $CaSO_4\, 2H_2O$

(c) $CaSO_4$

(d) $CaSO_4\, 1/2 H_2O$

Answer: (d)

6. Which oxide is amphoteric?

(a) BaO

(b) CaO

(c) BeO

(d) MgO

Answer: (c)

7. Be shows the diagonal relationship with

(a) Na

(b) Al

(c) Mg

(d) B

Answer: (b)

8. The tendency to lose their valance electron easily by alkali metals makes them

(a) strong reducing agent

(b) weak reducing agent

(c) strong oxidising agent

(d) weak oxidising agent

Answer: (a)

9. Which one is known as a fusion mixture?

(a) $Na_2CO_3 + NaHCO_3$

(b) $Na_2CO_3 + NaOH$

(c) $Na_2CO_3 + K_2SO_4$

(d) $Na_2CO_3 + K_2CO_3$

Answer: (d)

10. Find the incorrect trend for alkaline earth metals

(a) atomic size Be < Mg < Ca < Sr

(b) second ionization energy Be < Mg < Ca < Sr

(c) Hydration enthalpy Sr < Ca < Mg < Be

(d) Density Ca < Mg < Be < Sr

Answer: (b)

JEE

1. The correct statement for the molecule, CsI_3, is :

(1) it contains Cs^{3+} and I^- ions

(2) it contains Cs^+, I^- and lattice I_2 molecule

(3) it is a covalent molecule

(4) it contains Cs^+ and I_3^- ions.

Solution:

$CsI_3 \rightarrow Cs^+ + I_3^-$

CsI_3 contains Cs^+ and I_3^-.

Hence option (4) is the answer.

2. Which of the following statements about Na_2O_2 is not correct?

(1) Na_2O_2 oxidises Cr^{3+} to CrO_4^{2-} in acid medium

(2) It is diamagnetic in nature

(3) It is the super oxide of sodium

(4) It is a derivative of H_2O_2

Solution:

Na_2O_2 is sodium peroxide. It is not super oxide.

Hence option (3) is the answer.

3. The metal that cannot be obtained by electrolysis of

an aqueous solution of its salt is

1) Cr

2) Ag

3) Ca

4) Cu

Solution:

Ca cannot be obtained by the electrolysis of an aqueous solution of its salt.

Hence option (3) is the answer.

4. Based on lattice energy and other considerations, which one of the following alkali metal chloride is expected to have the highest melting point?

(1) RbCl

(2) LiCl

(3) KCl

(4) NaCl

Solution:

NaCl has the highest lattice energy and thus the highest melting point.

Hence option (4) is the answer.

5. A metal M on heating in nitrogen gas gives Y. Y on treatment with H_2O gives a colourless gas which when passed through $CuSO_4$ solution gives a blue colour, Y is:

(1) NH_3

(2) MgO

(3) Mg_3N_2

(4) $Mg(NO_3)_2$

Solution:

Magnesium, when heated with nitrogen, forms magnesium nitride.

3Mg + N2 → Mg3N2

Magnesium nitride reacts with water to form ammonia and magnesium hydroxide.

Mg3N2 → Mg(OH)2 + NH3

When ammonia gas is passed through CuSO4 solution, blue colour is obtained.

Hence option (3) is the answer.

6. Which one of the following will react most vigorously with water?

(1) Li

(2) K

(3) Rb

(4) Na

Solution:

Reactivity of s block elements increases down a group. So Rubidium will react most vigorously with water.

Hence option (3) is the answer.

7. Hydrogen bomb is based on the principle of

(1) Nuclear fission

(2) Natural radioactivity

(3) Nuclear fusion

(4) Artificial radioactivity

Solution:

Hydrogen bomb is based on the fusion of isotopes of hydrogen.

Hence option (3) is the answer.

8. Amongst LiCl, RbCl, BeCl$_2$ and MgCl$_2$ the compounds with the greatest and the least ionic character, respectively are :

(1) RbCl and MgCl$_2$

(2) LiCl and RbCl

(3) MgCl$_2$ and BeCl$_2$

(4) RbCl and BeCl$_2$

Solution:

Because of the smallest cationic size BeCl$_2$ is least ionic. Rb$^+$ has the biggest ionic size. So it has the greatest ionic character.

Hence option (4) is the answer.

9. In curing cement plasters water is sprinkled from time to time. This helps in

(1) developing interlocking needle-like crystals of

hydrated silicates.

(2) hydrating sand and gravel mixed with cement

(3) converting sand into silicic acid

(4) keeping it cool

Solution:

During hydration of calcium aluminosilicates, cross links are developed

Hence option (1) is the answer.

10. KO_2(potassium super oxide) is used in oxygen cylinders in space and submarines. This is because it

(1) absorbs CO_2 and increases O_2 content

(2) eliminates moisture

(3) absorbs CO_2

(4) produces ozone

Solution:

$2KO_2 + 2H_2O \longrightarrow 2KOH + H_2O_2 + O_2$. It eliminates moisture.

Hence option (2) is the answer.

11. A metal M readily forms its sulphate MSO_4 which is water soluble. It forms its oxide MO which becomes inert on heating. It forms an insoluble hydroxide $M(OH)_2$ which is soluble in NaOH solution. Then M is

(1) Mg

(2) Ba

(3) Ca

(4) Be

Solution:

Because of its small size, Beryllium shows anomalous properties with alkaline earth metals.

Hence option (4) is the answer.

12. The substance not likely to contain $CaCO_3$ is

(1) calcined gypsum

(2) sea shells

(3) dolomite

(4) a marble statue

Solution:

The composition of gypsum is $CaSO_4 \cdot 2H_2O$.

Hence option (1) is the answer.

13. Several blocks of magnesium are fixed to the bottom of a ship to

(1) make the ship lighter

(2) prevent action of water and salt

(3) prevent puncturing by under-sea rocks

(4) keep away the sharks

Solution:

Magnesium provides cathodic protection and prevents corrosion.

Hence option (2) is the answer.

14. The first ionization potential of Na is 5.1 eV. The value of electron gain enthalpy of Na+ will be

1) -2.55 eV

2) -5.1 eV

3) -10.2 eV

4) +2.55 eV

Solution:

Ionisation potential of alkali metal atom and electron affinity of its monocovalent cation are same in magnitude, but with opposite signs.

Hence option (2) is the answer.

15. In Zeolites and synthetic resin method, which will be more efficient in removing permanent hardness of water.

(1) Synthetic resin method as it exchanges only cation.

(2) zeolite resin method as it exchanges only cation.

(3) Synthetic resin method as it exchanges only anion.

(4) Synthetic resin is harmful to nature.

Solution:

The synthetic resin method will be more efficient in removing the permanent hardness of water because it exchanges only cation.

Hence option (1) is the answer.

P - BLOCK ELEMENTS

Question 1: Silicon dioxide is treated with hydrogen fluoride. Explain?

Answer 1: Silicon and hydrogen fluoride are treated to form silicon tetrafluoride. The bond is so strong that Halogens/Acids with Silicon dioxide do not allow the reaction. Still, they react at extreme temperatures, where the high electronegative fluorine substitutes silicon and Oxygen.

$$SiO_2 + 4HF \rightarrow SiO_4 + 2H_2O$$

Question 2. Why does silicon not form a graphite structure? Explain.

Answer 2: C is sp2 hybridised in graphite, and each C is linked to three other C atoms and forms hexagonal rings. This forms two- Dimensional sheet compositions of graphite.

Silicon does not form analogue C because of two reasons:

Silicon has a much lesser tendency for catenation than C as Si-Si bonds are much weaker than C-C bonds.
Because of its larger size than C, Silicon undergoes sp3 hybridisation.

Question 3. How is borax prepared from

(i) Colemanite ore (ii) Tincal (iii) Boric acid?

Answer 3:

(i) Colemanite ore: The mineral is boiled with sodium carbonate; as a result, they form perceptions of calcium carbonate, the crystal of borax and sodium metaborate.

$Ca_2B_6O_{11} + 2Na_2CO_3 \rightarrow Na_2B_4O_7 + 2NaBO_2 + 2CaCO_3$.

when mother liquid passes through carbon dioxide, it gets converted to borax

$4NaBO_2 + CO_2 \rightarrow Na_2B_4O_7 + Na_2CO_3$

(ii) Tincal: Naturally occurring borax is called Tincal; it is collected from dried lakes. We extract that Tincal from dried-up lakes, boil it with water, and filter it to remove impurities. The filter is then concentrated, and the crystal of borax separates.

(iii) Boric acid: Boric acid reacts with sodium carbonate; as a result, it forms a crystal of borax, water, and carbon dioxide.

$4H_3BO_3 + Na_2CO_3 \rightarrow Na_2B_4O_7 + 6H_2O + CO_2 \uparrow$

Question 4. Rationalise the Given Statements and Chemical Reactions:

Lead (II) chloride reacts with Cl_2 to give $PbCl_4$
Why Lead (IV) chloride is highly unstable toward heat.
Why Lead is known not to form an Iodide, PBI_4.
Answer (a): Due to the inert pair effect, PB shows an oxidation state of +2 and +4. Since Cl_2. is a strong oxidising agent, it oxidises to Pb_2 to Pb_4, and hence $PbCl_2$ reacts with $PbCl_4$.

$PbCl_2(s) + Cl_2(g) \rightarrow PbCl_4(l)$

Group +2 is more stable on moving down, and +4 is less stable due to the inert pair effect. Hence $pbcl_4$ is much more stable than $PbCl_2$. However, the formation of $PbCl_4$ takes place when Cl_2 gas is bubbled through a saturated solution of $PbCl_2$.

Answers (b): The greater stability of +2 over +4 oxidation state is because of the inert pair effect.

Lead (IV) chloride on heating decomposes to give Lead (II)

chloride and Cl2.

The reaction is given below:

PbCl4(l) → Δ PbCl2(s) + Cl2(g)

Answer (c): PB4 is oxidising in nature. A combination of PB4 and iodide is not stable. Iodide ions are strongly reduced in nature.

The reaction is given below:

Pb(IV) oxidises I- to I2 and itself gets reduced to PB(II)

PBI4 → PBI2 + I2

Question 5. Mention some important properties of carbon monoxide.

Answers 5. Some important properties of carbon monoxide are as follows:

It is a colourless, odourless gas, slightly soluble in water.
When carbon monoxide is treated with chlorine in the presence of light or charcoal, a poisonous gas is formed.
CO + Cl2 → COCl2

It burns in the air with a blue flame forming carbon dioxide
2CO + O2 → 2CO2.

Carbon monoxide acts as a strong agent when reacted with metals like iron, cobalt and nickel
Fe2O3 + 3CO → 2Fe + 3CO2.

Question 6: Explain the lower atomic radius of Gallium compared to Aluminium.

Answer 6: Aluminium size is 143 Pm, whereas Gallium size is 135 pm. Because of poor shielding of the Valence electron of Ga by the inner 3rd electron, the effective Nuclear charge of Ga is greater in magnitude than that of Al. As a result, the electron in Ga experiences a greater force of attraction by the Nucleus than in Al. Hence the atomic size of Ga is slightly less than that of Al.

Question 7: Explain the differences in properties of diamond and graphite based on their structures.

Answer 7: In a diamond, each carbon atom is sp3 hybridised and is bonded to four other carbon atoms through a sigma bond. In graphite, each carbon atom is sp2 hybridised and is bonded to three other carbon atoms through a sigma bond while the fourth electron forms a Π-bond.

The below table highlights the differences:

Sr no	Diamond	Graphite
1	It has crystalline lattice	It has a layered structure
2	It is made up of tetrahedral units	It has a planar geometry
3	The c-c bond of length in the diamond is 154 pm	The c-c bond length in graphite is 141.5 pm
4	They have a rigid covalent bond network, which is difficult to break	It is quite soft, and its layer can be represented easily.
5	It acts as an electrical insulator	It is a good conductor of electricity

Question 8 Explain why there is a phenomenal decrease in ionisation enthalpy from Carbon to silicon?

Answer 8: The ionisation enthalpy of Carbon (the first element of group 14) is very high (1086 KJ/mol). Moving down in a group increases its size with some screening effects. The force of attraction in the Nucleus for the Valence electron decreases too much. This results in a phenomenal decrease in ionisation enthalpy from Carbon to silicon.

Question 9: Why does boron trifluoride behave as a Lewis acid?

Answer 9: The B atom in Bf3 has only six electrons in the valence shell and hence is an electron-deficient molecule. They can easily accept a pair of electrons from Nucleophiles to complete their activities and thus behave as Lewis Acid.

Question 10: How is excessive CO2 responsible for global warming?

Answer 10: CO2 is an essential gas for our survival. CO2 is produced during combustion. Plants utilise it during photosynthesis, and O2 is released into the atmosphere. As a result of this CO2 cycle, a constant percentage of 21% O2 is maintained in the atmosphere. The decomposition of limestone and decrease in the number of trees has led to increased levels of CO2. which has the property of trapping the heat provided by sun rays. A higher amount of heat trapped increases the atmospheric temperature, causing global warming.

Question 11: Explain (i) Inert pair effect, (ii) Allotropy, and (iii) Catenation.

Answer 11:

(i) Inert pair effect: As we move down the group, the tendency of S – electrons of the valence shell to participate in a bond formation decreases. The reluctance of the S-electron to participate in bond formation is called the Inert-pair effect.

(ii) Allotropy: Allotropy is an element having more than one form. They have the same chemical properties but different physical properties.

For example, Carbon exists in three allotropic forms Diamond, Graphite and fullerenes.

(iii) Catenation: Atoms of some elements (such as Carbon) can link to strong covalent bonds to form long chains. This is the reason they are called catenation. They can be found in Carbon and are quite significant in Si and S.

Question 12: Why is CO poisonous? Explain.

Answer 12: In the lungs, Haemoglobin in Red Blood cells combines with molecular Oxygen loosely and forms oxyhaemoglobin.

Haemoglobin + Oxygen provides oxyhaemoglobin.

Co is highly poisonous because of the ability to form a complex molecule with haemoglobin. The Co-Hb complex is more stable (300 times) than the O_2-Hb complex.

Hb + CO → CO-Hb (carboxyhemoglobin)

As a result, the oxygen-carrying capacity of Hb is destroyed, and the person dies of suffocation as it displaces oxygen in the blood and deprives the heart, brain and other vital organs of oxygen.

Question 13: Explain the higher stability of BCl_3 as compared to $TICl_3$.

Answer 13: Due to poor shielding of the s-electron of the Valence shell (6s) by 3d, 4d, 5d and 4F electrons. The inert-pair effect is maximum in Tl; as a result, only $6p^1$ electrons participate in Bond formation, and thus the most stable state of Tl is +1 and not +3. Therefore TICl is stable, but $TICl_3$ is unstable.

In contrast, due to the absence of d and f electrons, B does not show an inert pair effect. In other words, all three valence electrons (i.e. two 2s and one 2p) participate in Bond formation. Hence, B shows an oxidation state of +3 and thus forms BCl_3. Thus BCl_3 is more stable than $TICl_3$.

COMPETITIVE CORNER

NEET

1. Find the amphoteric oxide

(a) CaO_2

(b) CO_2

(c) SnO_2

(d) SiO_2

Answer: (c)

2. Graphite has a structural similarity with

(a) B_2H_6

(b) B_4C

(c) B

(d) BN

Answer: (d)

3. Which is the correct order of decreasing acidity of lewis acids?

(a) $BBr_3 > BCl_3 > BF_3$

(b) $BF_3 > BCl_3 > BBr_3$

(c) $BCl_3 > BF_3 > BBr_3$

(d) $BBr_3 > BF_3 > BCl_3$

Answer: (a)

4. In the presence of KF, AlF_3 is soluble in HF. Find the complex formed

(a) $K_3[AlF_6]$

(b) AlH_3

(c) $K[AlF_3H]$

(d) $K_3[AlF_3H_3]$

Answer: (a)

5. S-S bond is present in which of the ion pairs

(a) $S_2O_7^{2-}, S_2O_3^{2-}$

(b) $S_4O_6^{2-}, S_2O_7^{2-}$

(c) $S_2O_7^{2-}, S_2O_8^{2-}$

(d) $S_4O_6^{2-}, S_2O_3^{2-}$

Answer: (d)

6. Which is the correct order of decreasing bond dissociation enthalpy?

(a) $F_2 > Cl_2 > Br_2 > I_2$

(b) $I_2 > Br_2 > Cl_2 > F_2$

(c) $Cl_2 > Br_2 > F_2 > I_2$

(d) $Br_2 > I_2 > F_2 > Cl_2$

Answer: (c)

7. Oxygen is not released on heating which of the compounds?

(a) $(NH_4)_2Cr_2O_7$

(b) $K_2Cr_2O_7$

(c) $Zn(ClO_3)_2$

(d) $KClO_3$

Answer: (a)

8. Which of the species has a permanent dipole moment?

(a) SF_4

(b) SiF_4

(c) BF_3

(d) XeF_4

Answer: (a)

9. Which of the statement is incorrect for XeO_4 ?

(a) four $p\pi$-$d\pi$ bonds are present

(b) four $sp^3 - p\pi$ bonds are present

(c) It has a tetrahedral shape

(d) It has a square planar shape

Answer: (d)

10. P_4O_{10} has _____ bridging O atoms

(a) 4

(b) 5

(c) 6

(d) 2

Answer: (c)

JEE

1. From the following statements regarding H_2O_2, choose the incorrect statement :

(1) It has to be stored in plastic or wax-lined glass bottles in the dark

(2) It has to be kept away from dust

(3) It can act only as an oxidizing agent

(4) It decomposes on exposure to light

Solution:

$Na_2SO_3 + H_2O_2 \rightarrow Na_2SO_4 + H_2O$

Hydrogen peroxide can act as both oxidising and reducing agent.

Hence option (3) is the answer.

2. The compound that does not produce nitrogen gas by the thermal decomposition is

(1) $(NH_4)_2Cr_2O_7$

(2) NH_4NO_7

(3) $(NH_4)_2SO_4$

(4) $Ba(N_3)_2$

Solution:

$(NH_4)_2SO_4 \rightarrow 2NH_3 + H_2SO_4$

On thermal decomposition of $(NH_4)_2SO_4$, NH_3 is evolved. On thermal decomposition of the other given compounds, N_2 is evolved.

Hence option (3) is the answer.

3. Glass is a

(1) micro-crystalline solid

(2) super-cooled liquid

(3) gel

(4) polymeric mixture.

Solution:

Glass is a transparent or translucent amorphous supercooled solid solution (supercooled liquid) of silicates and borates.

Hence option (2) is the answer.

4. Which one has the highest boiling point?

(1) Kr

(2) Xe

(3) He

(4) Ne

Solution:

The boiling point increases down the group.

Hence Xe has the highest boiling point in the given elements.

Hence option (2) is the answer.

5. Which of the following are Lewis acids?

(1) $AlCl_3$ and $SiCl_4$

(2) PH_3 and $SiCl_4$

(3) BCl_3 and $AlCl_3$

(4) PH_3 and BCl_3

Solution:

Lewis acid is an electron pair acceptor.

Both BCl_3 and $AlCl_3$ have vacant p- orbital and thus incomplete octet.

Hence they will act as Lewis acid.

Hence option (3) is the answer.

6. Which one of the following depletes the ozone layer?

(1) NO and freons

(2) SO_2

(3) CO

(4) CO_2

Solution:

NO and freons deplete the ozone layer.

Hence option (1) is the answer.

7. Which one of the following reactions of Xenon compounds is not feasible?

(1) $2XeF_2 + 2H_2O \rightarrow 2Xe + 4HF + O_2$

(2) $XeF_6 + RbF \rightarrow Rb[XeF_7]$

(3) $XeO_3 + 6HF \rightarrow XeF_6 + 3H_2O$

(4) $3XeF_4 + 6H_2O \rightarrow 2Xe + XeO_3 + 12HF + 1.5O_2$

Solution:

$XeF_6 + 3H_2O \rightarrow XeO_3 + 6HF$

XeF_6 has more tendency to hydrolyse. So the reverse reaction occurs.

Hence option (3) is the answer.

8. Assertion: Among the carbon allotropes, diamond is an insulator, whereas, graphite is a good conductor of electricity.

Reason: Hybridization of carbon in diamond and graphite are sp^3 and sp^2, respectively.

(1) Both assertion and reason are correct, but the reason is not the correct explanation for the assertion.

(2) Both assertion and reason are correct, and the reason is the correct explanation for the assertion.

(3) Both assertion and reason are incorrect.

(4) Assertion is an incorrect statement, but the reason is

correct.

Solution:

Diamond is a bad conductor of electricity due to the non-availability of free electrons. Graphite is a good conductor of electricity due to the fourth valence electron of each carbon which is free to move.

Hence option (1) is the answer.

9. Which of the following statements is wrong?

(1) Single N–N bond is weaker than the single P–P bond

(2) N_2O_4 has two resonance structures

(3) The stability of hydrides decreases from NH3 to BiH3 in group 15 of the periodic table

(4) Nitrogen cannot form d pi – p pi bond

Solution:

The stability of hydrides decreases from NH3 to BiH3 in group 15 of the periodic table.

Hence option (3) is the answer.

10. Boron cannot form which one of the following

anions?

(1) $B(OH)_4^-$

(2) BO_2^-

(3) BF_6^{3-}

(4) BH_4^-

Solution:

Because of the non-availability of d-orbitals, boron is unable to expand its octet. Hence the maximum covalence of boron cannot exceed 4.

Hence option (3) is the answer.

11. Which of the following statements regarding sulphur is incorrect?

(1) At $600°$ the gas mainly consists of S_2 molecules

(2) The oxidation state of sulphur is never less than +4 in its compounds

(3) S_2 molecule is paramagnetic

(4) The vapour at $200°$ C consists mostly of S_8 rings

Solution:

The oxidation state of sulphur is +2, +4, +6 and -2.

Hence option (2) is the answer.

12. The amorphous form of silica is

(1) cristobalite

(2) kieselguhr

(3) tridymite

(4) quartz.

Solution:

Kieselguhr is the amorphous form of silica.

Hence option (2) is the answer.

13. The gas evolved on heating CaF_2 and $SiO2$ with concentrated H_2SO_4, on hydrolysis gives a white gelatinous precipitate. The precipitate is:

(1) silica gel

(2) silicic acid

(3) hydrofluosilicic acid

(4) calcium fluorosilicate

Solution:

$H_2SO_4 + CaF_2 \rightarrow CaSO_4 + 2HF$

$4HF + SiO_2 \rightarrow SiF_4 + 2H_2O$

$3SiF_4 + 2H_2O \rightarrow 2H_2SiF_6 + SiO_2$

Hydrofluosilicic acid is H_2SiF_6

Hence option (3) is the answer.

14. Example of a three-dimensional silicate is :

(1) Beryls

(2) Zeolites

(3) Feldspars

(4) Ultramarines

Solution:

Feldspars are an example of three-dimensional silicate.

Hence option (3) is the answer.

15. Chlorine water on standing loses its colour and forms:-

(1) HCl and $HClO_2$

(2) HCl only

(3) HOCl and $HOCl_2$

(4) HCl and HOCl

Solution:

$Cl_2 + H_2O \rightarrow HCl + HOCl$

Hence option (4) is the answer.

16. What may be expected to happen when phosphine gas is mixed with chlorine gas?

(1) The mixture only cools down

(2) PCl_3 and HCl are formed and the mixture warms up

(3) PCl_5 and HCl are formed and the mixture cools down

(4) $PH_3 \cdot Cl_2$ is formed with warming up.

Solution:

$PH_3 + 4Cl_2 \rightarrow PCl_5 + 3HCl$

Phosphine gas when mixed with chlorine gas, gives phosphorus pentachloride and HCl.

Hence option (3) is the answer.

ORGANIC CHEMISTRY – SOME BASIC PRINCIPLES AND TECHNIQUES

Question 1. The reaction:

$CH_3CH_2I + KOH(aq) \rightarrow CH_3CH_2OH + KI$

is classified as :

(a) electrophilic substitution

(b) nucleophilic substitution

(c) elimination

(d) addition

Answer 1. (b) Nucleophilic substitution

Explanation:

It is a nucleophilic substitution reaction. As a nucleophile, the hydroxyl group of KOH with a lone pair of itself replaces the iodide ion in CH_3CH_2I to produce ethanol.

Question 2. Which of the following is the correct IUPAC name?

(i) 3-Ethyl-4, 4-dimethylheptane

(ii) 4,4-Dimethyl-3-ethylheptane

(iii) 5-Ethyl-4, 4-dimethylheptane

(iv) 4,4-Bis(methyl)-3-ethylheptane

Answer 2. (i) 3-Ethyl-4, 4-dimethylheptane

Explanation:

The various alkyl groups present in a compound are listed in the IUPAC name in alphabetical order. Di, tri, and tetra are left out of the alphabetical list. Therefore, the ethyl group (e) comes before the methyl group (m). As a result, the ethyl group is assigned the number 3 and is written first.

Question 3. Benzene hexachloride is prepared from benzene and chlorine in sunlight by

(a) Substitution reaction

(b) Elimination reaction

(c) Addition reaction

(d) Rearrangement reaction

Answer 3. (c) Addition reaction

Explanation: In addition to the reaction, benzene hexachloride is prepared from benzene and chlorine in sunlight. It occurs in the absence of oxygen, and a catalyst may be involved to increase the reaction rate.

$C_6H_6 + 3Cl_2 \rightarrow C_6H_6Cl_6$

Question 4. In the Lassaigne's test for nitrogen in an organic compound, the Prussian blue

colour is obtained due to the formation of:

(a) $Na_4[Fe(CN)_6]$

(b) $Fe_4[Fe(CN)_6]_3$

(c) $Fe_2[Fe(CN)_6]$

(d) $Fe_3[Fe(CN)_6]_4$

Answer 4. (b) $Fe_4[Fe(CN)_6]_3$

Explanation:

In Lassaigne's test for nitrogen in an organic molecule, the sodium fusion extract is heated with iron (II) sulphate. It is then acidified with sulphuric acid. Iron (II) sulphate and sodium cyanide first combine to create sodium hexacyanoferrate (II). Sulphuric acid is subsequently used to oxidise the iron (II), resulting in the production of iron (III)hexacyanoferrate (II) or Prussian blue.

The following are the chemical reactions:

$6CN^- + Fe^{2+} \longrightarrow [Fe(CN)_6]^{4-}$

$3[Fe(CN)_6]^{4-} + 4Fe^{3+} + xH_2O \longrightarrow Fe_4[Fe(CN)_6]_3 \cdot xH_2O$

(Prussian blue)

Question 5. The fragrance of flowers is due to the presence of some steam volatile organic

compounds called essential oils. These are generally insoluble in water at room temperature but are miscible with water vapour in the vapour phase. A suitable method for the extraction of these oils from the flowers is:

(i) Distillation

(ii) Crystallisation

(iii) Distillation under reduced pressure

(iv) Steam distillation

Answer 5. (iv) Steam distillation

Explanation:

Steam distillation is used to distinguish between compounds that are immiscible with water and are steam volatile. In steam distillation, heated flasks carrying the liquid to be distilled are passed through the steam produced by a generator. The steam and the volatile organic compound mixture are then condensed and collected. The separating funnel is afterwards used to separate the chemical from the water.

Question 6. Which of the following compounds will not exist as a resonance hybrid? Give reasons for your answer.

(i) CH_3OH

(ii) $R-CONH_2$

(iii) $CH_3CH=CHCH_2NH_2$

Answer 6. (i) CH_3OH and (iii) $CH_3CH=CHCH_2NH_2$

Explanation:

(i) CH_3OH does not contain pi-electrons and thus cannot form a resonance hybrid.

(iii) In $CH_3CH=CHCH_2NH_2$, the lone pair of electrons on the nitrogen atom is not coupled to the pi -electrons. Thus, the formation of a resonance hybrid is not possible.

Question 7. Which of the two:

$O_2NCH_2CH_2O-$ or CH_3CH_2O- is expected to be more stable, and why?

Answer 7. $O_2NCH_2CH_2O-$ is more stable.

Explanation:

Nitrogen being an electron-withdrawing group shows -I effect. It stabilises the molecule by lowering its negative electron charge. The methyl group, on the other hand, is an electron donor group and exhibits a +I effect. This makes the molecule more negatively charged and makes it less stable.

Question 8. In which of the following compounds, the Carbon marked with asterisk is

expected to have the greatest positive charge?

(i) *CH_3—CH_2—Cl

(ii) *CH_3—CH_2—Mg^+Cl^-

(iii) *CH_3—CH_2—Br

(iv) *CH_3—CH_2—CH_3

Answer 8. (i) *CH_3—CH_2—Cl

Explanation: According to the order of electronegativity:

Cl > Br > C > Mg.

The more electronegative group attached to the Carbon will give a more positive charge. Therefore, the C-atom having -Cl group attached to it will have the greatest positive charge.

Question 9. Which of the following compounds contain all the carbon atoms in the same

hybridisation state?

(i) H—C ≡ C—C ≡ C—H

(ii) CH_3—C ≡ C—CH_3

(iii) CH_2 = C = CH_2

(iv) CH_2 = CH—CH = CH_2

Answer 9. (i) H—C ≡ C—C ≡ C—H and (iv) CH_2 = CH—CH = CH_2

Explanation:

In these two compounds, all the carbon atoms have the same hybridisation state. In (i), all the carbon atoms are sp-hybridised, and in (iv), all the carbon atoms are sp2-hybridized.

Question 10. In the organic compound $CH_2 = CH - CH_2 - CH_2 - C \equiv CH$, the pair of hybridised

orbitals involved in the formation of the C2–C3 bond is:

(a) sp – sp2

(b) sp – sp3

(c) sp2 – sp3

(d) sp3 – sp3

Answer 10. (b) sp – sp3

Explanation:

In the given compound, the carbon atoms can be numbered as:

6   5   4   3   2   1

$CH_2 = CH - CH_2 - CH_2 - C \equiv CH$

The hybridisation of carbon atoms 1, 2, 3, 4, 5, and 6 results in sp, sp, sp3, sp3, sp2, and sp2 atoms, respectively. Therefore, the pair of hybridised orbitals involved in forming the C2-C3 bond is sp-sp3.

Question 11. Explain why alkyl groups act as electron donors when attached to a π system.

Answer 11. When connected to a -system, an alkyl group acts as an electron-donor group due to hyperconjugation. Take propane as an illustration.

Electron

Due to hyperconjugation, the sigma electrons of the C-H bond become delocalised. An unsaturated system is immediately joined to the alkyl. Delocalisation occurs when an sp3-s sigma bond orbital partially overlaps with an empty p orbital of a nearby carbon atom's n-bond.

orbital diagram showing hyperconjugation in propene

This kind of overlap causes the electrons to be delocalised, also known as no-bond resonance, which increases the stability of the molecule.

no-bond resonance

Question 12. Explain, how is the electronegativity of carbon atoms related to their state of

hybridization in an organic compound?

Answer 12. As the s-character grows, electronegativity rises. This is due to the nucleus's stronger attraction to s-electrons than p-electrons.

sp3 – 25% s-character, 75% p-character

sp2 – 33% s-character, 67% p-character

sp – 50% s-character, 50% p-character

Thus, the order of electronegativity is sp3 < sp2 < sp

Question 13. What are electrophiles and nucleophiles? Explain with examples.

Answer 13. A reagent known as an electrophile, or an electron-loving pair, requires an electron pair—for instance, carbonyl groups, CH3CH2+, C=O (due to the lone pair).

A reagent that contains an electron pair and is willing to contribute is known as a nucleophile. It is also referred to as a nucleus-loving reagent. These include NC–, OH–, and R3C– (carbanions).

Question 14. In DNA and RNA, nitrogen atoms are present in the ring system. Can Kjeldahl method be used for the estimation of nitrogen present in these? Give reasons.

Answer 14. No, the heterocyclic rings in DNA and RNA include nitrogen that cannot be eliminated as ammonia.

As nitrogen contained in these structures cannot be transformed into ammonium sulfate, the Kjeldahl method cannot be used to estimate nitrogen present in rings, azo, or nitro groups.

Question 15. Explain the principle of paper chromatography.

Answer 15. Partition chromatography includes paper chromatography. Chromatography paper is a unique kind of paper that is used in paper chromatography. Water that has been trapped inside a chromatography paper serves as the stationary phase.

The solution of the mixture is spotted at the base of a strip of chromatography paper primarily suspended in a suitable solvent or mixture of solvents. The mobile phase is this solvent. Capillary action causes the solvent to ascend up the paper and run over the spot. According to how they were divided differently into the two phases, the paper preserves particular components only in some cases. The produced paper strip is referred to as a chromatogram.

paper chromatography paper in two different shapes.

On the chromatogram, the spots representing the separated coloured compounds are visible at various heights from the location of the initial spot. Using a suitable spray reagent, as described under thin layer chromatography, or by using UV light, one can see the spots of the separated colourless chemicals.

Question 16. Give a brief description of the principles of the following techniques taking an example in each case.

(a) Crystallisation (b) Distillation (c) Chromatography

Answer 16.

Crystallisation:

Crystallisation is a method used to purify solid organic molecules. The difference in the solubility of the substance and impurities in a particular solvent is the basis on which it operates. Since the impure product is only sparingly soluble at lower temperatures, it is made to dissolve in the solvent at a higher temperature. We keep doing this until the solution is practically saturated. We obtain its crystals by cooling and filtering the mixture. For instance, we can obtain pure aspirin by crystallising 2-4g of crude aspirin in 20mL of ethyl alcohol. If necessary, it is heated and then left alone until it crystallises. After being separated, the crystals are dried.

Distillation:

Non-volatile liquids are separated from volatile components using this technique. Additionally, it is employed when the boiling temperatures of the constituents differ significantly.

It operates under the premise that liquids with various boiling points vaporise at various temperatures. The generated liquids are then separated after they have been cooled.

Ex: In a flask with a circular bottom and a condenser, aniline (b.p. = 457 K) and chloroform (b.p = 334 K) are combined. Due to its high volatility, chloroform vaporises first when heated and is then forced to pass through a condenser where it cools. The flask with a circular bottom still holds the aniline.

Chromatography:

It is popularly used for the separation and purification of

organic compounds.

It operates on the principle that each component of a mixture moves through the stationary phase under the influence of the mobile phase at a distinct rate.

Ex: A mixture of blue and red ink can be separated using chromatography. The component of the mixture less absorbed by the chromatogram is placed on the chromatogram. It causes the almost immobile component to travel more quickly up the paper than the other component.

Question 17. Give three points of differences between inductive effect and resonance effect.

Answer 17.

Inductive Effect	Resonance Effect
It involves the displacement of σ-electrons.	It involves the displacement of π-electrons or any lone pair of electrons.
This effect moves up to three carbon atoms before becoming insignificant after the fourth carbon atom.	The resonance effect moves all along the length of the conjugated system.
This operates in saturated compounds.	This operates only in unsaturated conjugated systems.

Question 18. Why does SO3 act as an electrophile?

Answer 18. The sulphur atom is attached to three highly electronegative oxygen atoms, resulting in the Sulphur atom becoming electron-deficient. Additionally, Sulphur also takes up a positive charge due to resonance. Both these factors make SO3 an electrophile.

electrophile

Question 19. The empirical formula of a compound is CH2. Its one mole has a mass of 42 g. What is its molecular formula?

Answer 19.

Molecular formula = n × empirical formula,

Therefore, Molecular formula = 4214 * CH2 = C3H6

Question 20. A mixture containing benzoic acid and nitrobenzene is given to you. Using an appropriate chemical reagent, how will you proceed to separate them?

Answer 20. When nitrobenzene reaches the organic layer, the mixture is agitated with a diluted NaHCO3 solution and extracted with ether or chloroform. The distillation of this will result in nitrobenzene. Dilute HCl is used to acidify the aqueous layer, and the solution is then cooled.

The final product is benzoic acid.

C6H5COOH + NaHCO3 → C6H5COONa + CO2 + H2O.

C6H5COONa + HCl (dilute) → C6H5COOH + NaCl.

Question 21. Discuss the chemistry of Lassaigne's test.

Answer 21. Lassaigne's test is used to identify the presence of phosphorus, halogens, nitrogen, and sulphur in an organic molecule. In an organic compound, these components are found in the covalent state. The chemical is fused with sodium metal to create the ionic form of these.

Lassaigne's

Boiling the fused mixture in distilled water releases the produced cyanide, sulfide, and sodium halide. The obtained extract is known as Lassaigne's extract. Then, the extract from Lassaigne is examined for the presence of phosphorus, sulphur, nitrogen, and halogens.

(a) Test for nitrogen

A Lassaigne test is performed on sodium fusion extract by heating it with iron (II) sulphate and then acidifying it with sulfuric acid. The first reaction involves sodium cyanide interacting with iron sulfate (II) to form sodium hexacyanoferrate (II). Then, some iron (II) is heated with sulphuric acid to create iron (III) hexacyanoferrate (II), which is Prussian blue. Here are the chemical equations involved in the reaction:

nitrogen

(b) Test for sulphur

Acetic acid is used to make the sodium fusion extract acidic to perform Lassaigne's test for sulphur in an organic compound. Next, lead acetate is added. Sulfur is present in the molecule because lead sulphide, which is black, precipitates out.

$S^{2+} + Pb^{2+} \rightarrow PbS$

(Black)

Sodium nitroprusside is used to treat the sodium fusion extract. The compound's appearance also indicates the presence of sulphur in the compound as violet.

$S^{2+} + [Fe(CN)_5NO]^{2-} \rightarrow [Fe(CN)_5NOS]^{4-}$

(Violet)

When both nitrogen and sulphur are present in an organic molecule, NaSCN is formed instead of NaCN.

$Na + C + N + S \rightarrow NaSCN$

$Fe^{3+} + SCN^- \rightarrow [Fe(SCN)]^{2+}$

(Blood red)

The colour of this NaSCN (sodium thiocyanate) is blood red. The lack of free cyanide ions prevents the formation of Prussian colour.

$NaSCN + 2Na \rightarrow NaCN + Na_2S$

(c) Test for halogens

To determine whether an organic molecule contains halogens, sodium fusion extracts are treated with silver nitrate after being acidified with nitric acid.

$X^- + Ag^+ \rightarrow AgX$

Here, X represents a halogen Cl, Br or I.

As a result of heating Lassaigne's extract, the nitrogen and sulphur are removed, which may interfere with the determination of halogens in organic compounds containing both elements.

COMPETITIVE CORNER

NEET

1. Which is the best-suited method for the separation of para and ortho-nitrophenols from 1:1 mixture?

(a) crystallisation

(b) chromatography

(c) sublimation

(d) steam distillation

Answer: (d)

2. Find the incorrect statement for a nucleophile

(a) A nucleophile is a Lewis acid

(b) Nucleophiles do not seek electron

(c) Ammonia is a nucleophile

(d) Nucleophiles attack low electron density sites

Answer: (a)

3. Which among the following is the most deactivating meta-directing group in aromatic substitution reaction?

(a) -COOH

(b) -SO_3H

(c) -NO_2

(d) -CN

Answer: (c)

4. Ammonia evolved from 0.75 g of the soil sample in the Kjeldahl's method for nitrogen estimation, neutralises 10 ml of 1M H_2SO_4. Find the percentage of nitrogen present in the soil

(a) 35.33

(b) 37.33

(c) 43.33

(d) 45.33

Answer: (b)

5. The correct order of increasing nucleophilicity is

(a) $Cl^- < Br^- < I^-$

(b) $Br^- < Cl^- < I^-$

(c) $I^- < Br^- < Cl^-$

(d) $I^- < Cl^- < Br^-$

Answer: (a)

6. Homologous series of alkanols have a general formula

(a) $C_nH_{2n}O_2$

(b) $C_nH_{2n}O$

(c) $C_nH_{2n+1}O$

(d) $C_nH_{2n+2}O$

Answer: (d)

7. Find the compound which undergoes nucleophilic

substitution reaction exclusively by an SN1 mechanism

(a) Benzyl chloride

(b) Chlorobenzene

(c) Ethyl chloride

(d) Isopropyl chloride

Answer: (a)

8. Which of the following methods is best suited for the separation of a mixture containing naphthalene and benzoic acid

(a) Crystallisation

(b) Chromatography

(c) Sublimation

(d) Distillation

Answer: (c)

9. How many structural isomers are possible if one hydrogen in diphenylmethane is replaced by chlorine?

(a) 8

(b) 4

(c) 7

(d) 6

Answer: (b)

10. Why do we boil the extract with conc. HNO_3 in Lassaigne's test for halogens?

(a) to increase the concentration of NO_3^- ions

(b) to increase the solubility product of AgCl

(c) it increases the precipitation of AgCl

(d) for the decomposition of Na_2S and NaCN formed

Answer: (d)

JEE

1. From the following statements regarding H_2O_2, choose the incorrect statement :

(1) It has to be stored in plastic or wax-lined glass bottles in the dark

(2) It has to be kept away from dust

(3) It can act only as an oxidizing agent

(4) It decomposes on exposure to light

Solution:

$Na_2SO_3 + H_2O_2 \rightarrow Na_2SO_4 + H_2O$

Hydrogen peroxide can act as both oxidising and reducing agent.

Hence option (3) is the answer.

2. The compound that does not produce nitrogen gas by the thermal decomposition is

(1) $(NH_4)_2Cr_2O_7$

(2) NH_4NO_7

(3) $(NH_4)_2SO_4$

(4) $Ba(N_3)_2$

Solution:

$(NH_4)_2SO_4 \rightarrow 2NH_3 + H_2SO_4$

On thermal decomposition of $(NH_4)_2SO_4$, NH_3 is evolved. On thermal decomposition of the other given compounds, N_2 is evolved.

Hence option (3) is the answer.

3. Glass is a

(1) micro-crystalline solid

(2) super-cooled liquid

(3) gel

(4) polymeric mixture.

Solution:

Glass is a transparent or translucent amorphous supercooled solid solution (supercooled liquid) of silicates and borates.

Hence option (2) is the answer.

4. Which one has the highest boiling point?

(1) Kr

(2) Xe

(3) He

(4) Ne

Solution:

The boiling point increases down the group.

Hence Xe has the highest boiling point in the given elements.

Hence option (2) is the answer.

5. Which of the following are Lewis acids?

(1) $AlCl_3$ and $SiCl_4$

(2) PH_3 and $SiCl_4$

(3) BCl_3 and $AlCl_3$

(4) PH_3 and BCl_3

Solution:

Lewis acid is an electron pair acceptor.

Both BCl_3 and $AlCl_3$ have vacant p- orbital and thus incomplete octet.

Hence they will act as Lewis acid.

Hence option (3) is the answer.

6. Which one of the following depletes the ozone layer?

(1) NO and freons

(2) SO_2

(3) CO

(4) CO_2

Solution:

NO and freons deplete the ozone layer.

Hence option (1) is the answer.

7. Which one of the following reactions of Xenon compounds is not feasible?

(1) $2XeF_2 + 2H_2O \rightarrow 2Xe + 4HF + O_2$

(2) $XeF_6 + RbF \rightarrow Rb[XeF_7]$

(3) $XeO_3 + 6HF \rightarrow XeF_6 + 3H_2O$

(4) $3XeF_4 + 6H_2O \rightarrow 2Xe + XeO_3 + 12HF + 1.5O_2$

Solution:

$XeF_6 + 3H_2O \rightarrow XeO_3 + 6HF$

XeF_6 has more tendency to hydrolyse. So the reverse reaction occurs.

Hence option (3) is the answer.

8. **Assertion:** Among the carbon allotropes, diamond is an insulator, whereas, graphite is a good conductor of electricity.

Reason: Hybridization of carbon in diamond and graphite are sp^3 and sp^2, respectively.

(1) Both assertion and reason are correct, but the reason is not the correct explanation for the assertion.

(2) Both assertion and reason are correct, and the reason is the correct explanation for the assertion.

(3) Both assertion and reason are incorrect.

(4) Assertion is an incorrect statement, but the reason is correct.

Solution:

Diamond is a bad conductor of electricity due to the non-availability of free electrons. Graphite is a good conductor of electricity due to the fourth valence electron of each carbon which is free to move.

Hence option (1) is the answer.

9. **Which of the following statements is wrong?**

(1) Single N–N bond is weaker than the single P–P bond

(2) N_2O_4 has two resonance structures

(3) The stability of hydrides decreases from NH3 to BiH3 in group 15 of the periodic table

(4) Nitrogen cannot form d pi – p pi bond

Solution:

The stability of hydrides decreases from NH3 to BiH3 in group 15 of the periodic table.

Hence option (3) is the answer.

10. Boron cannot form which one of the following anions?

(1) $B(OH)_4^-$

(2) BO_2^-

(3) BF_6^{3-}

(4) BH_4^-

Solution:

Because of the non-availability of d-orbitals, boron is unable to expand its octet. Hence the maximum covalence of boron cannot exceed 4.

Hence option (3) is the answer.

11. Which of the following statements regarding sulphur is incorrect?

(1) At 600^0 the gas mainly consists of S_2 molecules

(2) The oxidation state of sulphur is never less than +4 in its compounds

(3) S_2 molecule is paramagnetic

(4) The vapour at 200^0 C consists mostly of S_8 rings

Solution:

The oxidation state of sulphur is +2, +4, +6 and -2.

Hence option (2) is the answer.

12. The amorphous form of silica is

(1) cristobalite

(2) kieselguhr

(3) tridymite

(4) quartz.

Solution:

Kieselguhr is the amorphous form of silica.

Hence option (2) is the answer.

13. The gas evolved on heating CaF_2 and $SiO2$ with concentrated H_2SO_4, on hydrolysis gives a white gelatinous precipitate. The precipitate is:

(1) silica gel

(2) silicic acid

(3) hydrofluosilicic acid

(4) calcium fluorosilicate

Solution:

$H_2SO_4 + CaF_2 \rightarrow CaSO_4 + 2HF$

$4HF + SiO_2 \rightarrow SiF_4 + 2H_2O$

$3SiF_4 + 2H_2O \rightarrow 2H_2SiF_6 + SiO_2$

Hydrofluosilicic acid is H_2SiF_6

Hence option (3) is the answer.

14. Example of a three-dimensional silicate is :

(1) Beryls

(2) Zeolites

(3) Feldspars

(4) Ultramarines

Solution:

Feldspars are an example of three-dimensional silicate.

Hence option (3) is the answer.

15. Chlorine water on standing loses its colour and forms:-

(1) HCl and $HClO_2$

(2) HCl only

(3) $HOCl$ and $HOCl_2$

(4) HCl and $HOCl$

Solution:

$Cl_2 + H_2O \rightarrow HCl + HOCl$

Hence option (4) is the answer.

16. What may be expected to happen when phosphine

gas is mixed with chlorine gas?

(1) The mixture only cools down

(2) PCl_3 and HCl are formed and the mixture warms up

(3) PCl_5 and HCl are formed and the mixture cools down

(4) $PH_3 \cdot Cl_2$ is formed with warming up.

Solution:

$PH_3 + 4Cl_2 \rightarrow PCl_5 + 3HCl$

Phosphine gas when mixed with chlorine gas, gives phosphorus pentachloride and HCl.

Hence option (3) is the answer.

HYDROCARBONS

Ques: What is the structure of the alkene (C4H8) which is added on HBr with or without the presence of peroxide to produce the same product named C4H9Br.

Ans: 2-Butene that has the structure of CH3 – CH = CH — CH3 that is being symmetrical gives the same product, i.e., 2-bromobutane CH3 CH (Br) CH2CH3.

Ques: How to separate propene from propyne?

Ans: We can separate propene from propyne when the mixture is passed through ammoniacal AgNO3 solution where propyne reacts; on the other hand propene passes over.

Ques: Name the hydrocarbons in respect to the carbon-carbon bond.

Ans: with respect to the carbon–carbon bond that takes place between them, hydrocarbons are classified into three types:

(a) Saturated hydrocarbon

(b) Unsaturated hydrocarbon

(c) Aromatic hydrocarbon.

Ques: What are cycloalkanes?

Ans: Cycloalkanes are a type of saturated hydrocarbons that are formed when carbon atoms combine to form a closed chain or ring.

Ques: What do you understand by Lindlars' catalyst?

Ans: The Lindlars' catalyst is defined as partly deactivated palletized charcoal.

Ques: Write down the I.U.P.A.C. name of the alkane that has the lowest molecular weight and also contains a quaternary carbon.

Ans: 2, 2-Dimethylpropane.

Ques: Name three reagents that help to distinguish between ethyne and ethane.

Answer: Tollen's reagent, amm. CuCl solution and Ammoniacal $AgNO_3$.

Ques: What is the name of the simplest alkyne?

Answer: Ethyne

Ques: Cyclobutane is not as reactive as cyclopropane. Explain why?

Ans: Cyclobutane molecule has 90º C-C-C bond angles and in cyclopropane it has 60º. It explains that in cyclobutane, the bond angle difference between the tetrahedral and cyclopropane is smaller. On the other hand, when it is compared to cyclobutane, cyclopropane is majorly into severe strain and therefore they are more reactive in nature.

Ques: According to anti-markovnikov's rule, the peroxide effect is not visible in the case of HCl and HI that takes place in the presence of peroxide in addition to HBr to propene. Explain.

Ans: H—Br bonds are less stable than the H—Cl bond. Thus the H-Cl bonds are not divided by the free radical, whereas the H—I bond is weak in nature and iodine consists of free radicals that combine to produce iodine molecules in place of adding it to the double bond. Therefore no peroxide effect is seen while addition to HCl and HI.

Ques: During monochlorination of 2-methylpropane what is the new hydrocarbon radicals that are formed as intermediates? Which is the more stable product of them? Give reasons.

Ans: 2-methylpropane provides radicals of two types:

$CH_3-CH(CH_3)-*CH_2$
$CH_3-*C(CH_3)-CH_3$

Tertiary 3° free radical is a more stable compared to primary 1°, since it has 9α hydrogen, and the hyper-conjugation structure is also stabilised in nature. Primary 1° free radical is not a stable radical since it comprises only 1α hydrogen as well as only one hyper conjugative structure.

That is why (2) is more stable than (1)

Ques: How is alkyne produced from vinyl dihalides?

Ans: The preparation of alkynes from dihalides can be explained in two steps:

The process of dehydrohalogenation involves removing a halogen and a hydrogen atom which leads to the formation of vinylic halides.
The second step involves the halides that are made to react with a strong base and produce alkynes in the process. Large alkynes are made from the Small ones in the presence of Metal acetylides

Ques: How is cis – alkene formed by alkyne?

Ans: In the presence of palladised charcoal, alkyne are partially reduced with a distinct amount of di-hydrogen to form Cis-alkene and then they are partially deactivated using poisons like compounds such as sulphur compounds or quinoline.

Ques: How will you explain the structure of benzene?

Ans: In benzene, all the six carbon atoms are sp2 hybridized. Each carbon atom has two hybrid orbitals that overlap with the sp2 hybrid orbital of adjoining carbon atoms that create six C-C sigma bonds in the hexagonal plane structure. Now, each carbon atom stays back in the sp2 hybrid orbital that eventually intersects with the s-orbital of the hydrogen atom to give six C-H sigma bonds. For each carbon atom, only one hybridized p-orbital in a perpendicular position to the ring plane remains.

Carbon atoms that have unhybridized p-orbital will form the pi-bond.

Ques: Explain in brief the term polymerization with suitable examples.

Ans: When two or more molecules of unsaturated compounds combine to produce a larger complex under the given best conditions. The product that is produced aftermath is known as a polymer, and the process involved in it is called polymerization. Examples are:

(a) Additional polymerization: Nothing is wasted in this process as the larger molecule (polymer) is an exact multiple of the product produced from the smaller molecule.

(b) Condensation polymerization: During this process, molecules such as hydrochloric acid, water, and others are lost mostly. It does not even give an exact multiple of polymer produced from smaller molecules.

Ques: How will you differentiate between the following?
a) Butyne-1 & Butyne-2

Ans: Butyne-1 has an acetylene hydrogen atom which gives white ppt. when ammoniacal silver nitrate & red ppt. are combined with ammoniacal cuprous chloride. On the other hand, butyne-2- that have no acetylenic hydrogen atom that does not respond to any of the two reagents mentioned above.

(b) Butene-1 & Butene-2
To distinguish between Butene-1 & butene-2, two processes can be operated i.e. Ozonolysis or

Butene-1

Butene-1 & Butene-2

by oxidation with acidic KMnO2 solution with which they produce various carbonyl compounds.

Ques: What are the uses of acetylene?

Ans: Uses of acetylene are-

Welding and cutting is done by using Oxyacetylene flame.
It is also used as an illumining agent.
Artificial ripening of fruits is done with the help of Acetylene.
It is also used as a general anesthetic.

Ques: Why do addition reactions occur more often with alkenes & alkynes than with aromatic hydrocarbons?

Ans: When addition reaction occurs to an alkene or alkyne, the energy gained by forming two more sigma bonds makes up for the loss of the number of n bonds. However, in the case of aromatic hydrocarbons, the aromatic ring is firmly stabilized due to the delocalization of n electrons around the ring.

Hence it requires strong activation energy or force to claim the loss of its aromatic character. The most common reaction in arenes is the substitution reaction rather than addition reaction, as substitution does not lead to the loss of aromatic characteristics of the compound.

Ques: Explain in brief three processes involved in the process of preparation of alkynes.

Ans: The two processes involved in the preparation of alkynes are

Dehydrohalogenation: Alkanes formed after the reaction that takes place between alkenes and halogen thereby are passed through alcoholic KOH to form substituted alkenes. It then reacts with sodium amide which results in the formation of alkynes. This whole process is also called dehydrohalogenation as a hydrogen atom is replaced along with a halogen to get an alkyne in the result.

Preparation of Alkynes from Vicinal Dihalides: Vicinal dihalides prepare Alkynes with the help of the process of dehydrohalogenation. Dihalides are obtained from respective alkenes by adding halogen (group 17 elements) in the periodic table.

The laboratory preparation of alkynes can be explained in two steps; the first step comprises the preparation of unsaturated halides which are also known as vinylic halides. They are not very reactive in nature. Then this halide reacts to a strong base leading to the formation of alkynes which is denoted as the second step as well. Metal acetylides convert small alkynes into large ones in the process.

Preparation of Alkynes from Calcium Carbide: Calcium carbide is a major component that facilitates the process of alkynes synthesis on a larger scale. It is prepared by heating quicklime (CaO) in the presence of coke (C). The calcium carbide is later reacted with water, which leads to the evolution of calcium hydroxide and acetylene. The reaction is as follows:

$CaCO_3$
$\rightarrow$
$CaO + CO_2$

$CaO + 3C$
$\rightarrow$
$CaC_2 + CO$

CaC2 + 2H2O

→

Ca (OH)2 + C2H2

Ques: Write the formula involved in the conversion of the following compounds to benzene?

(i). Acetylene (ii). Cyclohexane (iii). Benzene diazonium chloride.

Answer: (i) Acetylene

When ethylene is heated to a higher temperature it polymerizes into benzene. The reaction is given below:

3C2H2

→

C6H6

(ii) Cyclohexane

Ans: When cyclohexane is exhibited to iron or quartz in a red hot tube, it starts to oxidise that leads to produce benzene. The reaction is given below:

Cyclohexane

Cyclohexane

(iii) Benzene diazonium chloride

Ans: Benzene diazonium chloride is converted to benzene in the presence of hypophosphorous acid. The reaction is given below:

Benzene diazonium chloride.

Benzene diazonium chloride

Ques: An alkyl halide C5H11Br (A) reacts to ethanolic KOH and produce an alkene 'B', which then reacts with Br2 to produce a compound 'C', which goes through dehydrobromination and gives an alkyne with 'D'. On treating with sodium metal in the presence of liquid ammonia, one mole of 'D' produces one mole of the sodium salt of 'D' and half a mole of hydrogen gas. Complete process of hydrogenation of 'D' produces a straight-chain alkane. Identify A, B, C and D from the given and also mention about the reactions involved.

Ans: Here the reaction implies that (D) is a terminal alkyne which means a triple bond is present at the end of the chain. It could be either (I) or (II).

Since alkyne 'D' when going through the process of hydrogenation gives out straight-chain alkane, the structure I is alkyne (D).

Hence, the structure of A, B and C are as follows:

CH3-CH2-CH2-CH2-CH2Br
CH3-CH2-CH2-CH=CH2
CH3-CH2-CH2-CH(Br)-CH2Br

COMPETITIVE CORNER

NEET CORNER:

Q: Identify product A in the following reaction : (NEET 2023)

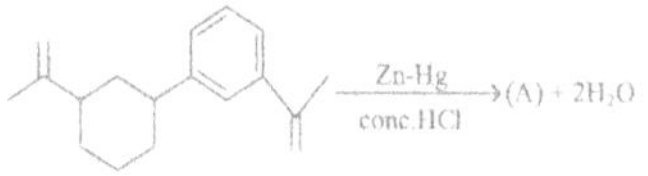

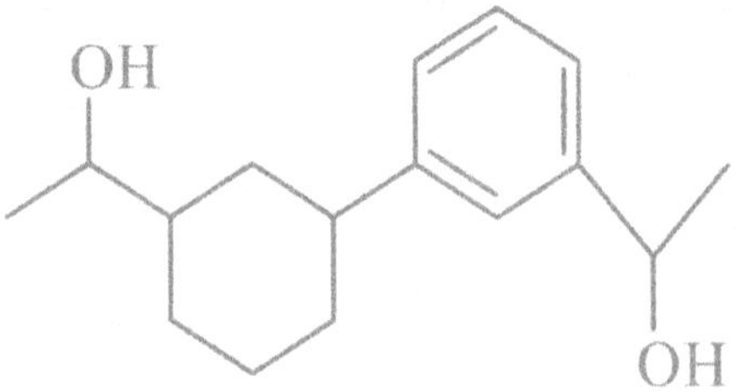

B. 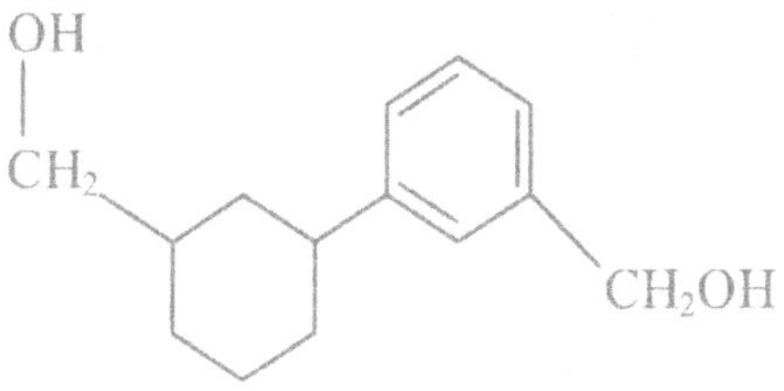

C. 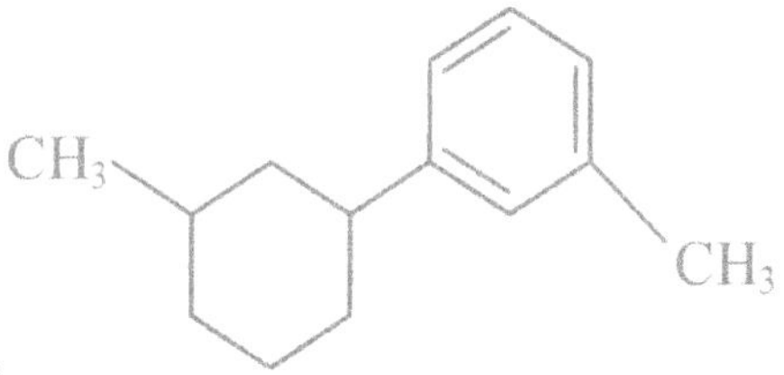

D. 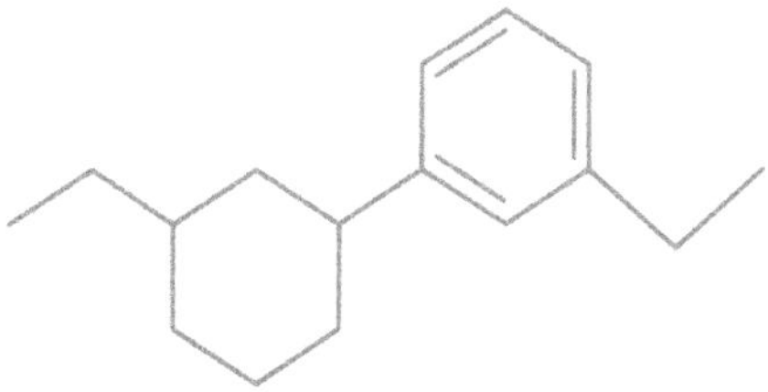

Answer : D

Q : Consider the following reaction and identify the product (P).

Q: Amongst the given options which of the following molecules/ion acts as a Lewis acid? (NEET 2023)

A.H2O
B.BF3
C.OH -
D.NH3

Answer : B

Q: The decreasing order of boiling points of the following alkanes is:

(a) heptane

(b) butane

(c) 2-methylbutane

(d) 2-methylpropane

(e) hexane

Q : Choose the correct answer from the options given below : (NEET 2022)

A. (a)>e>c>b>d
B. a>c>e>d>b
C. c>d>a>e>b
D. a>e>b>c>d

JEE CORNER

Q: The decreasing order of boiling points is [BHU 1999]

    A. N-Pentane > iso-Pentane > neo-Pentane
    B. iso-Pentane > n-Pentane > neo-Pentane
    C. neo-Pentane > iso-Pentane > n-Pentane
    D. n-Pentane > neo-Pentane > iso-Pentane

Q: Question 2. In the following reactions compound y is;

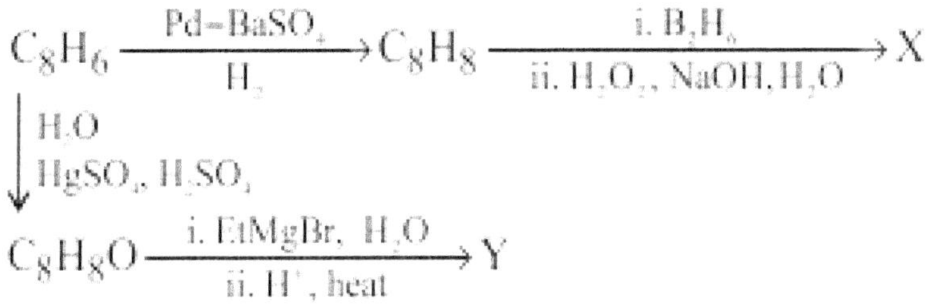

$C_8H_6 \xrightarrow[H_2]{Pd-BaSO_4} C_8H_8 \xrightarrow[\text{ii. H}_2\text{O}_2, \text{NaOH.H}_2\text{O}]{\text{i. B}_2\text{H}_6} X$

$\downarrow$ H$_2$O, HgSO$_4$, H$_2$SO$_4$

$C_8H_8O \xrightarrow[\text{ii. H}^+, \text{heat}]{\text{i. EtMgBr, H}_2\text{O}} Y$

(a) 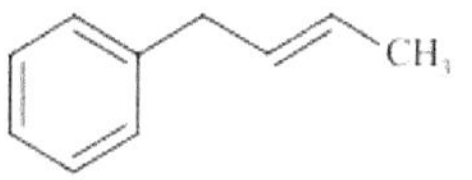 Ph-CH$_2$-CH=CH-CH$_3$

(b) 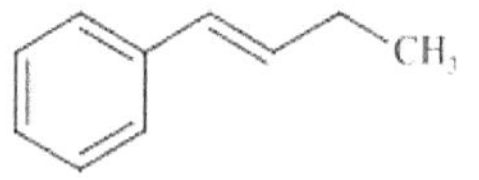 Ph-CH=CH-CH$_2$-CH$_3$

(c) Ph-C(=CH$_2$)-CH$_2$-CH$_3$

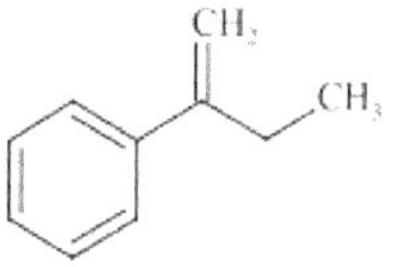

(d) 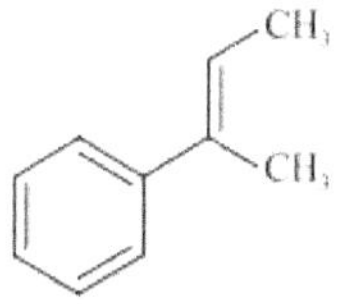 Ph-C(CH$_3$)=CH-CH$_3$

ANS D

Q: The bond energy (in kcal mol^{-1}) of a C–C single bond is approximately:

A. 1

B. 10

C. 100

D. 1000

ANS : C

Q: The total number of stereoisomers that can exist for M is:

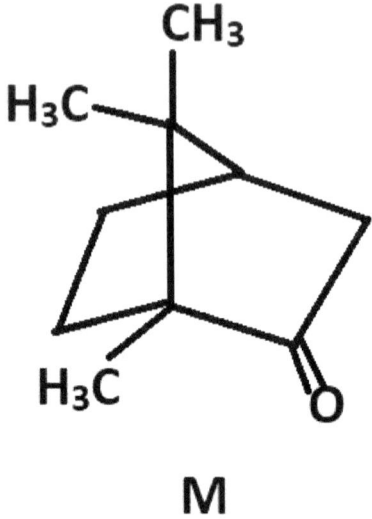

M

ANS : 2

Q: Newman projections P, Q, R and S are shown below:

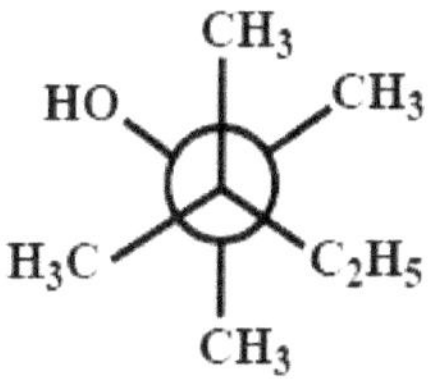

Which one of the following options represents identical molecules?

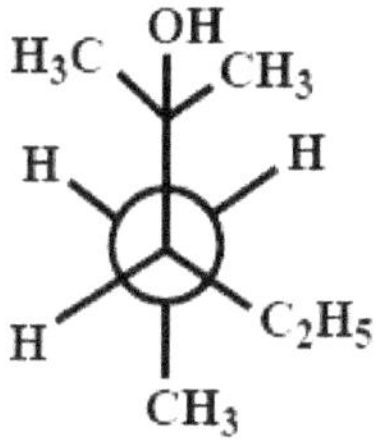

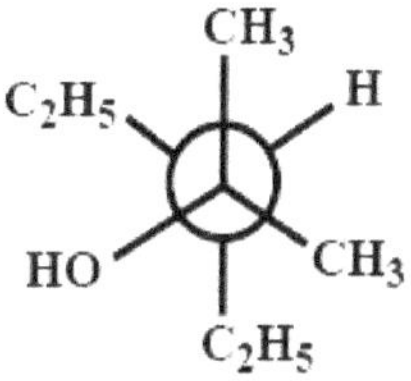

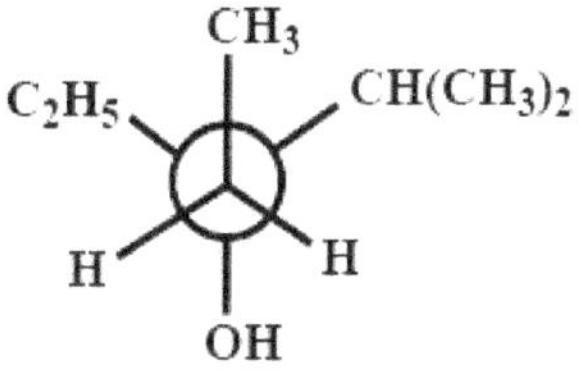

A. P and Q

B. Q and S

C. Q and R

D. R and S

Solution: (C)

On monochlorination of 2-methylbutane, the total number of chiral compounds obtained is.

A. 2

B. 4

C. 6

D. 8

Solution: (A)

Q: The correct statement(s) for the following addition reactions is(are):

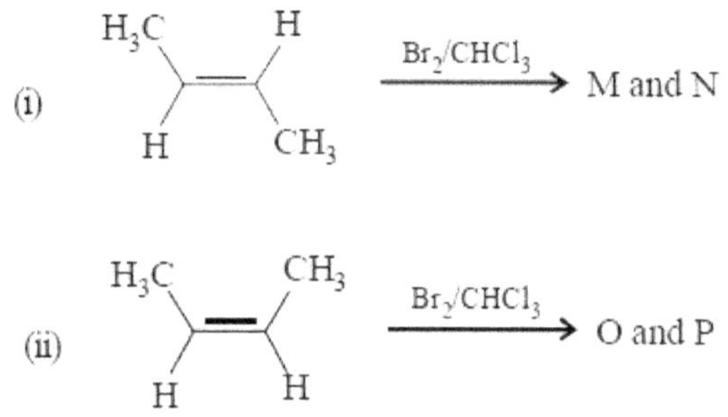

(i) H_3C, H on one carbon; H, CH_3 on other (cis-2-butene) $\xrightarrow{Br_2/CHCl_3}$ M and N

(ii) H_3C, CH_3 on one carbon; H, H on other (trans-2-butene) $\xrightarrow{Br_2/CHCl_3}$ O and P

A. O and P are identical molecules

B. (M and O) and (N and P) are two pairs of diastereomers

C. (M and O) and (N and P) are two pairs of enantiomers

D. Bromination proceeds through trans-addition in both the reactions

Solution: (B and D)

1. The most basic compound among the following is:-

(1) Acetanilide

(2) Benzylamine

(3) p-Nitro aniline

(4) Aniline

Solution:

Basicity is inversely proportional to resonance of lone pair electrons. Benzylamine is more basic. The electron pairs do not involve in resonance in benzylamine. In other amines, there is delocalization of lone pair of electron on N atom on the ring. In acetanilide, the delocalization of lone pair of electrons on N atom is due to adjacent CO group.

Hence option (2) is the answer.

2. Considering the basic strength of amines in aqueous solution, which one has

the smallest pK_b value?

(1) $(CH_3)_3N$

(2) $C_6H_5NH_2$

(3) $(CH_3)_2NH$

(4) CH_3NH_2

Solution:

Because of resonance, aryalamines are less basic than alkyl amines. In aryalamines, the lone pair of electrons on N is partly shared with the ring and is less available for sharing with a proton. In alkylamines, the electron releasing alkyl group increases the electron density on nitrogen atom. Hence increases the protonation ability of amines. So, more the number of alkyl groups, higher will be basicity of amine. Because of steric effect, a slight discrepancy occurs in case of trimethyl amine. So $(CH_3)_2NH$ has the smallest pK_b value.

Hence option (3) is the answer.

3. Among the following the molecule with the lowest dipole moment is:-

(1) $CHCl_3$

(2) CH_2Cl_2

(3) CCl_4

(4) CH_3Cl

Solution:

Dipole moment of CCl_4 is 0.

The order of dipole moment is CCl_4 < CH_3Cl < CH_2Cl_2 < $CHCl_3$.

Hence option (3) is the answer.

4. Which one of the following compounds will not be soluble in sodium bicarbonate ?

(1) Benzene sulphonic acid

(2) Benzoic acid

(3) o-Nitrophenol

(4) 2, 4, 6 – Trinitrophenol

Solution:

Benzene sulphonic acid and Benzoic acid are stronger acids and they react with sodium bicarbonate. o-Nitrophenol is a very weak acid and it does not react with sodium bicarbonate. o-Nitrophenol will not be soluble in sodium bicarbonate. 2, 3, 6 – Trinitrophenol is higher in acidity and it reacts with sodium bicarbonate.

Hence option (3) is the answer.

5. The compound formed in the positive test for nitrogen with the Lassaigne solution of an organic compound is

(1) $Fe_4[Fe(CN)_6]_3$

(2) $Na_4[Fe(CN)_5NOS]$

(3) $Fe(CN)_3$

(4) $Na_3[Fe(CN)_6]$

Solution:

The compound formed in the positive test for nitrogen with the Lassaigne solution of an organic compound is prussian blue coloured complex compound ferric ferrocyanide.

Hence option (1) is the answer.

6. The general formula $C_nH_{2n}O_2$ could be for open chain

(1) carboxylic acids

(2) diols

(3) dialdehydes

(4) diketones

Solution:

The general formula $C_nH_{2n}O_2$ could be for open chain carboxylic acid or ester.

Hence option (1) is the answer.

7. Among the following oxoacids, the correct decreasing order of acid strength is :

(1) $HClO_4 > HClO_3 > HClO_2 > HOCl$

(2) $HClO_2 > HClO_4 > HClO_3 > HOCl$

(3) $HOCl > HClO_2 > HClO_3 > HClO_4$

(4) $HClO_4 > HOCl > HClO_2 > HClO_3$

Solution:

Acidic strength is directly proportional to the oxidation number. Increasing acid strength is because of an increase in the oxidation state of the central atom.

Hence option (1) is the answer.

8. Ortho-Nitrophenol is less soluble in water than p- and m- Nitrophenols because:

(1) Melting point of o–Nitrophenol is lower than those of m– and p– isomers

(2) o–Nitrophenol is more volatile in steam than those of m– and p– isomers

(3) o–Nitrophenol shows Intramolecular H–bonding

(4) o–Nitrophenol shows Intermolecular H–bonding

Solution:

Intramolecular H-bonding is present in o-nitrophenol. So solubility in water is decreased.

Hence option (3) is the answer.

9. The order of basicity of amines in gaseous state is :

(1) $3^0 > 2^0 > NH_3 > 1^0$

(2) $1^0 > 2^0 > 3^0 > NH_3$

(3) $NH_3 > 1^0 > 2^0 > 3^0$

(4) $3^0 > 2^0 > 1^0 > NH_3$

Solution:

The basicity is proportional to +I effect. The presence of electron-donating group increases the basicity of amines. The presence of withdrawing group decreases the basicity of amines.

Hence option (4) is the answer.

10. Which one of the following conformation of cyclohexane is chiral ?

(1) Twist boat

(2) Rigid

(3) Chair

(4) Boat

Solution:

Twist boat is chiral because it does not have a plane of symmetry.

Hence option (1) is the answer.

11. Increasing order of stability among the three main conformations (i.e. Eclipse, Anti, Gauche) of 2-

fluoroethanol is

(1) Eclipse, Gauche, Anti

(2) Gauche, Eclipse, Anti

(3) Eclipse, Anti, Gauche

(4) Anti, Gauche, Eclipse

Solution:

Eclipse is least stable and Gauche is most stable.

Hence option (3) is the answer.

12. The conjugate base of hydrazoic acid is :-

(1) NH_3^-

(2) N_3^-

(3) N_2^-

(4) N^{-3}

Solution:

Hydraulic acid is HN_3.

$HN_3 \rightarrow H^+ + N_3^-$

Hence option (2) is the answer.

13. The correct order of increasing basicity of the given conjugate base (R = CH$_3$) is:

(1) RCOO$^-$ < HC ≡ C$^-$ < NH$_2^-$ < R$^-$

(2) RCOO$^-$ < HC ≡ C$^-$ < R$^-$ < NH$_2$

(3) R$^-$ < HC ≡ C$^-$ < RCOO$^-$ < NH$_2^-$

(4) RCOO < NH$_2$ < HC ≡ C$^-$ < R$^-$

Solution:

The basic strength is inversely proportional to the stability of conjugate base. In basicity, if the availability of the electrons is more, then more readily they can be donated to form a new bond to the proton and, and hence the stronger base. If bronsted acid is a strong acid then its conjugate base is a weak base.

The correct order of increasing basic strength is RCOO$^-$ < HC ≡ C$^-$ < NH$_2^-$ < R$^-$

Hence option (1) is the answer.

14. Which types of isomerism is shown by 2,3-dichlorobutane?

(1) Diastereo

(2) Optical

(3) Geometric

(4) Structural

Solution:

2,3- dichlorobutane shows optical isomerism.

Hence option (2) is the answer.

15. Due to the presence of an unpaired electron, free radicals are

(1) Chemically reactive

(2) Chemically inactive

(3) Anions

(4) Cations

Solution:

Due to the presence of an unpaired electron, the free radicals are chemically active.

Hence option (1) is the answer.

16. Which one the following does not have

sp^2 hybridized carbon?

(1) Acetone

(2) Acetamide

(3) Acetonitrile

(4) Acetic acid

Solution:

Acetonitrile is having only sp^3 and sp hybridized carbon atoms.

Hence option (3) is the answer.

ENVIRONMENTAL CHEMISTRY

Question 1: Dinitrogen and dioxygen are the main constituents of air but these do not react with each other to form oxides of nitrogen because _____.

(i) the reaction is endothermic and requires a very high temperature.

(ii) the reaction can be initiated only in presence of a catalyst.

(iii) oxides of nitrogen are unstable.

(iv) N2 and O2 are unreactive.

Answer 1: (i)

Explanation: Due to the presence of a triple bond and the extremely high dissociation energy of N2, these gases do not react with one another at room temperature.

Question 2: Which of the following statements about photochemical smog is wrong?

(i) It has a high concentration of oxidising agents.

(ii) It has a low concentration of the oxidising agent.

(iii) It can be controlled by controlling the release of NO2, hydrocarbons, ozone etc.

(iv) Plantation of some plants like pinus helps in controlling photochemical smog.

Answer 2: (ii)

Explanation: Photochemical smog contains a lot of oxidants as it contains formaldehyde, nitric oxide, peroxyacetyl nitrate, and acrolein. Photochemical smog has many negative health effects.

Question 3: Which of the following statements is wrong?

(i) Ozone is not responsible for the greenhouse effect.

(ii) Ozone can oxidise sulphur dioxide present in the atmosphere to sulphur

trioxide.

(iii) Ozone hole is thinning of the ozone layer present in the stratosphere.

(iv) Ozone is produced in the upper stratosphere by the action of UV rays on

oxygen.

Answer 3: (i)

Explanation: O_3 makes up between 8 and 10% of the greenhouse effect. The earth's surface absorbs about 75% of solar energy, while the remaining 25% is reflected back into space. This heat traps atmospheric gases like CO_2, CH_4, O_3, CFCs, and H_2O, increasing the temperature of the atmosphere and contributing to global warming.

Question 4: Which of the following statements is not true about classical smog?

(i) Its main components are produced by the action of sunlight on emissions of automobiles and factories.

(ii) Produced in a cold and humid climates.

(iii) It contains compounds of reducing nature.

(iv) It contains smoke, fog and sulphur dioxide.

Answer 4: (i)

Explanation: Due to the gases that are generated by factories and automobiles, classic smog develops in cold, humid climates. The classic smog is made up of a mixture of SO_2, smoke, and fog.

Question 5: Which of the following gases is not a greenhouse gas?

(i) CO

(ii) O_3

(iii) CH_4

(iv) H_2O vapour

Answer 5: (i)

Explanation: Carbon monoxide (CO) is not considered a direct greenhouse gas, mostly because it does not absorb terrestrial thermal IR energy strongly enough

Question 6: Define Environmental Chemistry.

Answer 6: Environmental Chemistry is the study of chemical and biological processes that take place in the natural environment. It also explores the interaction, origin, effects, and movement of biochemical species on earth.

Question 7: What is the troposphere?

Answer 7: The lowest part of the atmosphere is the

troposphere. It is the layer of the atmosphere where people and other animals can be found. It rises to a height of roughly 10 km above sea level.

Question 8: List gases which are responsible for the greenhouse effect.

Answer 8: The main cause of the greenhouse effect is carbon dioxide(CO_2). Methane(CH_4), nitrous oxide(NO), water vapour(H_2O), CFCs, and ozone(O_3) are other greenhouse gases.

Question 9: Which disease is caused due to ozone layer depletion?

Answer 9: When the ozone layer in the atmosphere is depleted, the chance of skin cancer among living organisms increases. Through the ozone layer's gaps, the sun's UV radiation penetrates the earth and causes ailments to the skin.

Question 10: What is the desired concentration of Fluoride ion (F^-) in drinking water?

Answer 10: PHS advises community water systems that add fluoride to their water to reduce the incidence of dental fluorosis to have a fluoride concentration of 0.7 mg/L (parts per million [ppm]).

Question 11: What do you mean by Biochemical Oxygen Demand (BOD)?

Answer 11: Biochemical Oxygen Demand is the volume of oxygen that bacteria need to consume in order to break down the organic material that is present in a specific amount of the water sample. A BOD level of less than 5 ppm indicates that the water is clean, whereas one of 17 ppm indicates that the water is extremely polluted.

Question 12: What is PAN?

Answer 12: Peroxyacetyl nitrate or PAN is a secondary

pollutant that is present in photochemical smog. When heated, it breaks down into peroxy ethanol radicals and nitrogen dioxide gas.

Question 13: What are biodegradable and non-biodegradable pollutants?

Answer 13: Biodegradable pollutants are those that can be broken down by bacteria. It includes waste from fruits and vegetables, cow dung, and other organic materials.

Non-biodegradable pollutants are those that cannot be broken down by microbes. It includes substances like mercury, polythene, DDT, and others.

Question 14: How can domestic waste be used as manure?

Answer 14: The garbage must first be divided into biodegradable and non-biodegradable segments. Food wastes and other materials that can be broken down by bacteria are biodegradable and are disposed of in landfills with the microorganisms that break them down. The final product, Humus, a degraded substance, can be used as manure in crops. The remainder of the garbage, which cannot be decomposed, must be recycled.

Question 15: Why does water cover with excessive algal growth become polluted?

Answer 15: The addition of fertilisers with phosphate increases the growth of algae on the water's surface, which makes the water unfit for swimming or boating and gives off a foul odour. It also reduces the amount of oxygen in the water, which can be dangerous for aquatic life.

Question 16: What is pneumoconiosis?

Answer 16: A pneumoconiosis is a form of interstitial lung illness brought on by breathing specific kinds of lung-damaging dust particles. Being more likely to be exposed to these clouds of dust at work, pneumoconiosis is considered to be occupational lung disease.

Question 17: Carbon monoxide gas is more dangerous than carbon dioxide gas. Why?

Answer 17: Carbon dioxide (CO2) and carbon monoxide (CO) gases are released when different fuels are burned. While carbon dioxide is naturally non-toxic, carbon monoxide is harmful.

Carbon monoxide is dangerous because it can combine with haemoglobin to produce a more stable complex than the oxygen-haemoglobin complex, carboxyhemoglobin. The blood's ability to carry oxygen is reduced when carboxyhemoglobin levels are between 3 and 4%. Headaches, poor vision, jitters, and cardiovascular issues are common occurrences. At higher concentrations, it can be more fatal. In normal cases, carbon dioxide is not dangerous but it can prove to be fatal at higher concentrations.

Question 18: The greenhouse effect leads to global warming. Which substances are responsible for the greenhouse effect?

Answer 18: There are many different greenhouse gases, some of which are produced naturally and others that are created by humans. Natural gas called methane is created when vegetation burns, digests, or rots in the absence of oxygen. Large amounts of methane are released into the atmosphere by paddy fields, coal mines, fossil fuels, and decaying waste.

Although nitrous oxide is present in the atmosphere naturally, it is rising daily as a result of human activity.

ACs employ CFCs (Chlorofluorocarbons), which are man-made. Additionally, they produce greenhouse gases.

Question 19: What is the importance of measuring the BOD of a water body?

Answer 19: BOD is a measurement of the amount of organic material in water based on how much oxygen is needed for biological breakdown. Clean water is defined as having a BOD value of less than 5 ppm, whereas severely contaminated water has a BOD value of 17 ppm or above.

Question 20: What do you mean by green chemistry? How will it help decrease environmental pollution?

Answer 20: The goal of green chemistry is to develop and execute chemical products and processes that will limit the use and synthesis of compounds that are harmful to the environment by utilising the chemistry that is already known and understood and its underlying principles.

Environmental pollution results from the emission of several hazardous compounds (particulates, gases, organic wastes, and inorganic wastes). In green chemistry, the reactants that will be used in chemical reactions are selected so that the final products will produce up to 100% of their total yield. This minimises or inhibits the release of chemical contaminants into the environment. Tetrachlorethane and chlorine gas have been substituted with H_2O_2 in the drying and bleaching of paper due to the efforts of green chemists.

Question 21: What are the harmful effects of oxides of nitrogen in the atmosphere?

Answer 21: The harmful effects of oxides of nitrogen in the atmosphere are as follows:

High amounts of NO_2 injure plants by causing leaf spotting, a reduction in photosynthetic activity, and suppression of vegetative growth.
In humans, nitric oxide leads to bronchitis and respiratory problems. It results in photochemical haze and acid rain.
Nitrogen oxide fractures rubber and harms fibres such as nylon, rayon, and cotton.

They also interact with ozone in the atmosphere, reducing ozone density.

Question 22: Ozone is a toxic gas and is a strong oxidising agent. Even then its presence in the stratosphere is very important. Explain what would happen if ozone from this region is completely removed.

Answer 22: Ozone gas, which makes up the stratosphere, shields us from the sun's dangerous UV rays (= 225 mm). UV radiation can injure people and result in melanoma, eye cataracts, genetic mutations, and crop devastation. Aquatic creatures and plants may also have a negative impact. The atmospheric release of chlorofluorocarbons (CFCs) is the primary cause of ozone depletion.

So, even though O_3 is detrimental to us while it is in the troposphere, it shields us from radiation when it is in the stratosphere.

Question 23: Statues and monuments in India are affected by acid rain. How?

Answer 23: Acid rain is a consequence of a number of human activities that release sulphate and nitrogen oxides into the atmosphere. After being oxidised, these oxides combine with water vapour to create acids.

$2SO_2(g) + O_2(g) + 2H_2O(l) \rightarrow 2H_2SO_4(aq)$

$4NO_2(g) + O_2(g) + 2H_2O(l) \rightarrow 4HNO_3(aq)$

Buildings and infrastructures composed of stone and metal suffer are damaged from acid rain. A significant stone utilised in the Taj Mahal and other famous statues and structures in India is limestone.

Limestone responds to acid rain as follows:

$CaCO_3 + H_2SO_4 \rightarrow CaSO_4 + H_2O + CO_2$

Thus, the monuments become pale and lose their colour and lustre.

Question 24: Ozone is a gas heavier than air. Why does the ozone layer not settle down near the earth?

Answer 24: There is a dynamic equilibrium between the creation and decomposition of ozone because it is a thermodynamically unstable gas and can be broken down into molecular oxygen.

$O_2(g) \rightarrow O(g) + O(g)$

$O(g) + O_2(g) \rightarrow O_3(g)$

Question 25: What are the reactions involved for ozone layer depletion in the atmosphere?

Answer 25: The reactions involved in ozone layer depletion in the atmosphere are:

CFCs are discharged into the atmosphere and combined with other gases before reaching the stratosphere, where UV light decomposes them.

$CF_2CL_2(g) \rightarrow Cl+(g) + CF_2Cl(g)$

The chlorine-free radical produced in the first steps reacts with the ozone as:

$Cl-(g) + O_3(g) \rightarrow ClO-(g) + O_2(g)$

The chlorine free radical further reacts with atomic oxygen to produce more chlorine free radicals as follows:

$ClO-(g) + [O] \rightarrow Cl-(g) + O_2(g)$

Question 26: Explain tropospheric pollution in 100 words.

Answer 26: Tropospheric pollution is mostly caused by the presence of unfavourable elements in the atmosphere's base layer, such as solid or gaseous particles.

The following are the principal pollutants found in the troposphere:

Gaseous Pollutants: These mostly include sulphur (SO_2 & SO_3), nitrogen, and carbon oxides, hydrogen sulphide(H_2S), hydrocarbons, ozone, and other oxidants.

Particulate pollutants: The principal components include smog, dust, mist, and fumes.

Burning fossil fuels such as coal and gasoline releases sulphur and nitrogen oxides, which when they come into contact with water make nitric acid (HNO_3) and sulphuric acid(H_2SO_4), respectively, and so cause acid rain.

$$2SO_2 + O_2 + H_2O \rightarrow 2H_2SO_4$$

$$4NO_2 + O_2 + 2H_2O \rightarrow 4HNO_3$$

Question 27: Green plants use carbon dioxide for photosynthesis and return oxygen to the atmosphere, even then carbon dioxide is considered to be responsible for the greenhouse effect. Explain why.

Answer 27: CO_2 contributes to global warming. Although it is caused by the combustion of fossil fuels, plants take in CO_2 for photosynthesis and release oxygen in return, which slows down global warming.

Only found in the troposphere, CO_2 makes up 0.03% of the total volume of the atmosphere.

But as we all know, using fossil fuels and deforestation both result in higher CO_2 levels, which further adds up to global warming.

Question 28: Some time ago the formation of polar stratospheric clouds was reported over Antarctica. Why were these formed? What happens when such clouds break up by the warmth of sunlight?

Answer 28: Over the South Pole of Antarctica, scientists working there noticed ozone layer depletion or the existence of an ozone hole. It was found that the ozone hole was caused by a certain set of components based on the seasons.

While polar stratospheric clouds build up over Antarctica in the winter, nitrogen dioxide and methane combine with chlorine monoxide and chlorine atoms in the summer to create chlorine sinks that stop ozone depletion.

During the winter, polar stratospheric clouds above Antarctica provide a surface on which hypochlorous acid is formed by the hydrolysis of chlorine nitrate. It also reacts with HCl to produce molecular chlorine.

$ClO + NO_2 (g) \rightarrow ClONO_2 (g)$

$Cl (g) + CH_4 \rightarrow CH_3 (g) + HCl (g)$

$ClONO_2 (g) + H_2O \rightarrow HOCl (g) + HNO_3$

$ClONO_2 (g) + HCl \rightarrow Cl_2 (g) + HNO_3 (g)$

In the spring, as the sun rises once again, the warmth of the sun penetrates the cloud, protolyzing HOCl and Cl2.

$HOCl (g) \rightarrow OH(g)Cl(g)$

$Cl_2 \rightarrow 2Cl (g)$

The resulting chlorine radicals trigger the chain of ozone depletion.

Question 29: Write down the reactions involved during the formation of photochemical smog.

Answer 29: The interaction of sunlight with hydrocarbons and nitrogen oxides results in photochemical smog. Common elements of photochemical smog include ozone, nitric oxide, acrolein, formaldehyde, and peroxyacetyl nitrate (PAN). The process that causes photochemical smog to occur can be summed up as follows:

The burning of fossil fuels releases nitrogen dioxide and hydrocarbons into the atmosphere. High levels of these pollutants in the air cause them to interact with sunlight in the following ways:

$NO_2(g) \rightarrow NO(g) + O(g)$

Nitrogen dioxide   Nitric oxide

$O(g) + O_2(g) \leftrightarrow O_3(g)$

$O_3(g) + NO(g) \rightarrow NO_2(g) + O_2(g)$

While both NO_2 and O_3 are oxidising agents, ozone is hazardous by nature. They create formaldehyde, PAN, and acrolein when they interact with the unburned hydrocarbons in the air.

$3CH_4 + 2O_3 \rightarrow 3CH_2=O + 3H_2O$

Formaldehyde

Question 30: Explain how does greenhouse effect cause global warming.

Answer 30: The term "greenhouse effect" describes global warming that occurs as a result of the atmosphere absorbing heat that is radiated from Earth into space. Some gases in the atmosphere act like the glass in a greenhouse, letting sunlight enter but preventing the heat from the Earth from escaping into space. Gases including water vapour, carbon dioxide (CO_2), methane, nitrous oxides, and chlorofluorocarbons all contribute to the greenhouse effect (CFCs).

COMPETITIVE CORNER

NEET

1. Photochemical smog normally does not contain

(a) Chlorofluorocarbons

(b) Peroxyacetyl nitrate

(c) Ozone

(d) Acrolein

Answer: (a)

2. Depletion of the ozone layer is caused due to

(a) Ferrocene

(b) Fullerenes

(c) Freons

(d) Polyhalogens

Answer: (c)

3. Find the incorrect statement

(a) BOD value of clean water is less than 5 ppm

(b) Drinking water pH should be between 5.5-9.5

(c) carbon, sulphur and nitrogen oxides are the most widespread air pollutants

(d) dissolved oxygen concentration below 5 ppm is ideal for the growth of fish

Answer: (d)

4. Find the secondary pollutant among these

(a) PAN

(b) N_2O

(c) SO_2

(d) CO_2

Answer: (a)

5. The reaction responsible for the radiant energy of the Sun is

(a) nuclear fission

(b) nuclear fusion

(c) chemical reaction

(d) combustion

Answer: (b)

6. Alum's capacity to purify water is due to

(a) softens hard water

(b) pathogenic bacteria get destroyed

(c) impurities' coagulation

(d) it improves taste

Answer: (c)

7. The coldest region of the atmosphere

(a) Troposphere

(b) Thermosphere

(c) Stratosphere

(d) Mesosphere

Answer: (d)

8. Which of the oxide of nitrogen is not a common pollutant?

(a) N_2O_5

(b) N_2O

(c) NO

(d) NO_2

Answer: (a)

9. The compound essential for the process of photosynthesis has this element

(a) Ca

(b) Ba

(c) Fe

(d) Mg

Answer: (d)

10. In the air, N_2 and O_2 occur naturally but they do not react to form oxides of nitrogen because

(a) oxides of nitrogen are unstable

(b) catalyst is required for the reaction

(c) the reaction is endothermic

(d) N_2 and O_2 do not react with each other

Answer: (c)

11. This about carbon monoxide is incorrect.

(a) It is produced due to incomplete combustion

(b) The carboxyhaemoglobin (haemoglobin found to CO) is less stable than oxyhaemoglobin

(c) It reduces the oxygen-carrying ability of blood

(d) It forms carboxyhaemoglobin

Answer: (b)

12. **This is a sink for CO**

(a) Haemoglobin

(b) Oceans

(c) Micro organisms present in the soil

(d) Plants

Answer: (c)

13. **DDT is**

(a) Nitrogen containing insecticide

(b) Biodegradable pollutant

(c) Non-Biodegradable pollutant

(d) An antibiotic

Answer: (c)

14. Which of the following techniques is/are used in controlling water pollution?

(a) Reverse osmosis

(b) Ion exchange process

(c) Adsorption process

(d) All of these

Answer: (d)

15. Which of the following pollutants cannot be degraded by natural process?

(a) Heavy metals

(b) DDT

(c) Nuclear waste

(d) All of these

Answer: (d)

16. A substance which may alter environmental constituents or cause pollution is referred to as

(a) Pollutant

(b) Decomposer

(c) Radiator

(d) Reducer

Answer: (a)

17. Regular use of which of the following fertilizers increases the acidity of soil?

(a) Potassium nitrate

(b) Superphosphate of lime

(c) Ammonium sulphate

(d) Urea

Answer: (c)

18. Which of the following metal will pollute water?

(a) Cd

(b) Na

(c) K

(d) None of the above

Answer: (a)

JEE

1. Which is wrong with respect to our responsibility as a human being to protect our environment?

(a) Avoiding the use of floodlighted facilities

(b) Setting up compost tin in gardens

(c) Using plastic bags

(d) Restricting the use of vehicles

Solution:

Using plastic bags should be avoided to protect the environment.

Hence option (c) is the answer.

2. BOD stands for

(a) Biochemical Oxidation Demand

(b) Biological Oxygen Demand

(c) Biochemical Oxygen Demand

(d) Bacterial Oxidation Demand.

Solution:

BOD stands for Biochemical Oxygen Demand.

Hence option (c) is the answer.

3. The layer of atmosphere between 10 km and 50 km above the sea level is called

(a) thermosphere

(b) mesosphere

(c) stratosphere

(d) troposphere.

Solution:

The stratosphere is the layer of atmosphere between 10 km and 50 km above the sea level.

Hence option (c) is the answer.

4. Addition of phosphate fertilizers to water bodies causes

(a) enhanced growth of algae

(b) increase in amount of dissolved oxygen in water

(c) deposition of calcium phosphate

(d) increase in fish population.

Solution:

The addition of phosphate fertilizers to water bodies causes enhanced growth of algae.

Hence option (a) is the answer.

5. Water samples with BOD values of 4 ppm and 18 ppm, respectively, are

(a) clean and highly polluted

(b) highly polluted and highly polluted

(c) highly polluted and clean

(d) clean and clean.

Solution:

Clean water has a BOD value less than 5 ppm. Highly polluted water has BOD value of 17 ppm or more.

Hence option (a) is the answer.

6. Excessive release of CO_2 into the atmosphere results in

(a) global warming

(b) formation of smog

(c) polar vortex

(d) depletion of ozone.

Solution:

Excessive release of CO_2 into the atmosphere results in global warming.

Hence option (a) is the answer.

7. The higher concentration of which gas in air can cause stiffness of flower buds?

(a) SO_2

(b) CO

(c) NO_2

(d) CO_2

Solution:

The higher concentration of SO_2 in the air can cause stiffness of flower buds.

Hence option (a) is the answer.

CHEMISTRY PART 1

8. Air pollution that occurs in sunlight is

(a) oxidising smog

(b) fog

(c) reducing smog

(d) acid rain.

Solution:

The main components of the photochemical smog result from the action of sunlight on unsaturated hydrocarbons and nitrogen oxides produced by factories and automobiles. Photochemical smog has a high concentration of oxidising agent. So, it is called oxidising agent.

Hence option (a) is the answer.

9. The gas leaked from a storage tank of the Union Carbide plant in Bhopal gas tragedy was

(a) phosgene

(b) methylisocyanate

(c) methylamine

(d) ammonia.

Solution:

Methylisocyanate was the reason for the Bhopal gas tragedy.

Hence option (b) is the answer.

10. The maximum prescribed concentration of copper in drinking water is

(a) 0.05 ppm

(b) 3 ppm

(c) 5 ppm

(d) 0.5 ppm

Solution:

The maximum prescribed concentration of copper in drinking water is 3 ppm.

Hence option (b) is the answer.

11. Taj Mahal is being slowly disfigured and discoloured. This is primarily due to

(a) acid rain

(b) soil pollution

(c) water pollution

(d) global warming

Solution:

Acid rain is the reason for the discolouration of Taj Mahal.

Hence option (a) is the answer.

12. The upper stratosphere consisting of the ozone layer protects us from the sun's radiation that falls in the wavelength region of

(a) 400-550 nm

(b) 600-750 nm

(c) 200-315 nm

(d) 0.8-1.5 nm

Solution:

Ozone protects us from the sun's radiation that falls in the wavelength region of 200-315 nm.

Hence option (c) is the answer.

13. Identify the wrong statement in the following.

(a) Acid rain is mostly because of the oxides of nitrogen and sulphur.

(b) Chlorofluorocarbons are responsible for ozone layer depletion.

(c) Greenhouse effect is responsible for global warming.

(d) Ozone layer does not permit infrared radiation from the sun to reach the earth.

Solution:

Ozoneplanket is the thick layer of ozone which is effective in absorbing harmful ultraviolet rays given out by the sun. It acts as a protective shield. It does not permit the ultraviolet rays from the sun to reach the earth.

Hence option (d) is the answer.

14. Which one of the following substances used in dry cleaning is a better strategy to control environmental pollution?

(a) Sulphur dioxide

(b) Carbon dioxide

(c) Nitrogen dioxide

(d) Tetrachloroethylene

Solution:

Liquid carbon dioxide is better to replace conventional halogenated solvents (potentially carcinogenic).

Hence option (b) is the answer.

15. The smog is essentially caused by the presence of

(a) O_2 and O_3

(b) O_2 and N_2

(c) oxides of sulphur and nitrogen

(d) O_3 and N_2.

Solution:

Smog is caused by oxides of sulphur and nitrogen.

Hence option (c) is the answer.

16. Which of the following conditions in drinking water causes methemoglobinemia?

(a) > 50 ppm of chloride

(b) > 50 ppm of nitrate

(c) > 50 ppm of lead

(d) > 100 ppm of sulphate

Solution:

The concentration of nitrate greater than 50 ppm in drinking water causes methemoglobinemia.

Hence option (b) is the answer.

BOOKS IN THIS SERIES

CLASS 11

Physics Part 1

www.ingramcontent.com/pod-product-compliance
Lightning Source LLC
Chambersburg PA
CBHW052340220526
45465CB00003BA/887